# Practical
# English
# Handbook

As part of Houghton Mifflin's ongoing commitment to the environment, this text has been printed on recycled paper.

# Houghton Mifflin Company ▪ Boston

*Geneva, Illinois*　　　　*Palo Alto*　　　　*Princeton, New Jersey*

# Practical English Handbook

TENTH EDITION

**Floyd C. Watkins**
*Emeritus, Emory University*

**William B. Dillingham**
*Emory University*

Authors of works quoted in the text are identified with the quotations.

Grateful acknowledgment is made to the following publishers and authors for permission to reprint from their works.

Diane Ackerman, "The Real George Washington." Reprinted by permission from *Parade,* copyright © 1988.

Ron Carlson, "On the U.S.S. Fortitude," copyright © by Ron Carlson. First appeared in *The New Yorker.* Copyright © 1990 by Ron Carlson. Reprinted by permission of W. W. Norton & Co.

William Gaylin, "Conduct and the Inner Self." Copyright © 1977 by *The New York Times Company.* Reprinted by permission.

Denise Levertov, "In Mind." Poems 1960–1967. Copyright © 1964 by Denise Levertov. Reprinted by permission of *New Directions Publishing Corporation,* and *Laurence Pollinger Limited.*

Douglas L. Wilson, Copyright © 1991, Douglas L. Wilson. Used by permission of the author, "What Jefferson and Lincoln Read," first appeared in *The Atlantic Monthly,* January 1991.

Cover designer: Clifford Stoltze, Stoltze Design, Boston
Cover image: Michael Goldman / FPG International

Printed in the U.S.A.

Library of Congress Catalog Card Number 95-76995

Student's Edition ISBN: 0-395-73333-2

Instructor's Annotated Edition ISBN: 0-395-73334-0

3456789-QK-99 98 97 96

# Contents

# Sentence Errors    29

# Punctuation   **123**

## Mechanics   173

## Literature    325

## Research    353

# Preface

With each new edition of the *Practical English Handbook,* its authors have aimed at retaining the simplicity, directness, and usefulness of the original version while addressing the changing needs of students as they learn and of instructors as they teach. During the preparation of this Tenth Edition, we have examined all aspects of the book, and we have applied to everything the test of practicality, which we interpret as true helpfulness. This scrutiny has resulted in numerous improvements, a partial list of which is included below. We have

- refurbished illustrative sentences with an eye toward affording students a quicker understanding of the problems to be identified and corrected;

- adjusted the level of difficulty in exercises so that they now contain more realistic problems and are more readily accessible to the average student of today;

- streamlined and clarified precepts throughout the book, most notably in the discussion of verbs and verb tenses, which has been reorganized to make precepts clearer and more direct, and in the two sections dealing with agreement of subject and verb and of pronoun and antecedent;

- included a new section that addresses specifically and exclusively the proper use of *who* and *whom;*

- created helpful new sections in the treatment of adjectives and adverbs that deal with absolute concepts *(unique, perfect)* and with avoiding double comparatives and superlatives and with double negatives;

- reorganized several parts of the book, such as the discussions of choppy sentences and excessive coordination, separation of elements, and variety in sentences, so that subsections are now devoted to separate issues;

- added a new section to the discussion of sentence variety that offers guidelines for varying the openings of sentences;

- simplified terminology throughout, for example, *essential* for *restrictive* and *nonessential* for *nonrestrictive;*

- expanded the treatment of unnecessary commas to include comma errors not previously covered;

- here and there added exercises where users have requested them, for example, an exercise on the proper use of colons;

- included new material on preparing manuscripts on a word processor and added a section on using the spell checker on a computer;

- provided instructions on preparing a resume and furnished a model;

- reorganized the chapter on diction for greater effectiveness and created a section dealing with the error of mistaking one word for another;

- supplied all new vocabulary exercises;

- added a section on exaggeration to the discussion of logic and accuracy;

- repositioned the chapter on writing papers to come after that on writing paragraphs for a more logical organization;

- revised extensively the discussion of writing paragraphs to make it more coherent and useful;

- added a new section on "Writing About Yourself" together with a model paper for the chapter on writing papers;

- expanded the introduction to the chapter on writing about literature for increased insight into the difference between the literary essay and other types of papers;
- added to the chapter on writing about literature new subsections that deal with the proper ways to begin a paper and with the use of quotations as evidence;
- provided a model paper on a poem by Denise Levertov;
- brought up to date the chapter on writing the research paper by
  a. updating lists of general reference aids and periodical indexes available in libraries,
  b. expanding explanations of online library catalogs and including illustrations of computer displays,
  c. adding a discussion of abstracts available on CD-ROM with an illustration;

- replaced the previous model paper on dolphins with a new paper on the subject of magazine sweepstakes;
- added exercises in the chapter on writing the research paper to help the student understand the MLA method of documentation;
- altered examples of APA documentation to reflect changes necessitated by the most recent edition of the *Manual of the Amercian Psychological Association,* 4th ed., 1994;
- set off in a distinctive manner the section on APA documentation to make it readily distinguishable from the MLA system;
- added a brief but practical section on English as a Second Language (ESL);
- expanded the Glossary of Usage to include words and phrases that have recently become controversial;
- broadened the Glossary of Grammatical Terms (now the Glossary of Terms) to include definitions not only of grammatical terms but also of many other terms useful to students of writing.

To our countless colleagues in the effort to teach the basics of good writing, we give thanks that must of necessity be feeble compared to our debt. You and the authors of this book are linked by common experience and by a common goal. Many of you have told us honestly and openly what you want in a handbook; others have frankly and constructively criticized; many have offered invaluable suggestions. It is difficult to imagine how a handbook could evolve through ten editions without your help. Most of you cannot be listed by name in a brief preface, but to all of you we owe a continuing debt. Professor Susan VanZanten Gallagher of Seattle Pacific University rewrote several sections of the Tenth Edition. Because of her special talents, the book is richer. Eric and Marie Nitschke of the Woodruff Library, Emory University, offered expert advice for the section on research tools. Various members of the Department of English, Emory University, helped with suggestions and support. We also express our gratitude to the following for their assistance.

Landrum Banks, Xavier University of Louisiana

Dick Bollenbacher, Edison State Community College, OH

John Bonner, Ricks College, ID

Joe Cox, United States Military Academy, West Point, NY

Natalie Collura Farina, Lehman College and City University of New York

Billye Givens, Eastern Oklahoma State College

John Gregg, San Diego Mesa College, CA

Robert Scattergood, Belmont Technical College, OH

Terri Shanahan, Grand Valley State University, MI

Floyd C. Watkins
William B. Dillingham

# Grammar

Anyone who uses language, who puts words together to communicate ideas, knows something about grammar. The word *grammar* refers to the often unconscious principles that guide people as they use language. It also can refer to the formal study of a grammatical system and its rules. For example, the person who says "They are" is using grammar. A person who says *"They* is a plural pronoun and therefore takes the plural verb *are"* is using grammatical terms and rules to explain how Standard English works. The technical study of grammar can be conducted in great detail, but this book presents only basic elements of the grammatical system—what you need in order to communicate effectively in writing and speech.

## The Parts of Speech

A basic aspect of grammar is knowing the parts of speech. This gives you a vocabulary for identifying words and talking about how language works.

The eight parts of speech are **nouns, pronouns, verbs, adjectives, adverbs, conjunctions, prepositions,** and **interjections.** Each of these is explained below.

The function of a word within a sentence determines what part of speech that word is. For example, a word may be a **noun** in one sentence but an **adjective** in another.

*noun*
↓
She teaches in a *college.*

*adjective*
↓
She teaches several *college* courses.

## Nouns

**Nouns** are words that name. They also have various forms that indicate **gender** (masculine, feminine, neuter), **number** (singular, plural), and **case** (see **Glossary of Terms**). Nouns are classified as follows:

(a) **Proper nouns** name specific groups, people, places, or things. They usually begin with a capital letter *(Toni Morrison, London, Knicks)*.

   *Thomas Jefferson* built *Monticello* in *Virginia.*

(b) **Common nouns** name general groups, people, places, or things *(reader, city, players)*.

   Few *authors* write anonymously.

(c) **Collective nouns** name a whole group but are singular in form *(navy, family, choir)*.

   The *jury* is weighing the evidence.

(d) **Abstract nouns** name concepts, beliefs, or qualities that cannot be perceived with the five senses *( faith, honor, joy)*.

   Her *love* of *freedom* was as obvious as her *courage.*

(e) **Concrete nouns** name things that can be perceived with the five senses *(rain, desk, heat)*.

   *Steam* rose from the simmering *stew.*

## Pronouns

Most **pronouns** take the place of a noun. Some pronouns (such as *something, none, anyone*) have general or broad references and do not directly take the place of a particular noun.

Pronouns fall into categories that both classify how they stand for nouns and how they function in a sentence. (Some words that function as pronouns can also have other uses; that is, they sometimes function as a different part of speech.)

(a) **Personal pronouns** usually refer to a person, sometimes to a thing. They have many forms, which depend on their grammatical function.

|  | SINGULAR | PLURAL |
|---|---|---|
| First person | I, me, mine | we, us, ours |
| Second person | you, yours, | you, yours |
| Third person | he, she, it | they, them, theirs |
|  | his, hers, its |  |

(b) **Indefinite pronouns** do not refer to a particular person or thing. Some of the most common are *some, any, each, everyone, everybody, anyone, anybody, one, neither* (see **8f**).

*indefinite pronoun*
↓
*Everyone* likes praise.

(c) **Reflexive pronouns** end in *-self* or *-selves* and indicate that the subject acts upon itself.

*reflexive pronouns*
↓
I hurt *myself*.
You should remember to protect *yourself*.

(d) **Intensive pronouns** also end in *-self* or *-selves*. An intensive pronoun emphasizes a noun that precedes it in the sentence.

*intensive pronoun*
↓
I *myself* will carry the message.

*intensive pronoun*
↓

Julie *herself* will eat the frozen yogurt.

(e) **Demonstrative pronouns** point to specific objects or people (see **demonstrative adjectives,** p. 8). There are four demonstrative pronouns: *this* and *that* are singular; *these* and *those* are plural.

*demonstrative pronoun*
↓

Many varieties of apples are grown here. *These* are winesaps.

(f) **Interrogative pronouns** are used in asking questions: *what, who, whom, whose, which* (see **9f**).

*interrogative pronoun*
↓

*Who* was chosen?

(g) **Relative pronouns** *(who, whoever, whom, whomever, that, what, which, whose)* generally introduce dependent adjective clauses (see p. 25) and refer back to noun or pronoun antecedents in independent clauses. They serve as connectives between dependent and independent clauses.

┌─ *adjective clause* ─┐
│  *relative pronoun*  │
↓

Special rooms are provided for guests *who* need quiet rest.

Relative pronouns also introduce noun clauses (see p. 25) and in this function do not refer back to a definite noun or pronoun.

┌─ *noun clause* ─┐
│ *relative pronoun* │
↓

*Whoever* wins friends will have great riches.

# ■ Exercise 1

*Identify nouns and pronouns.*

1. Amelia Earhart was one of the most famous aviators in the thirties; almost everyone in the United States could identify her.

2. George Palmer Putnam was a brash publicist who managed Earhart's career.

3. Earhart herself was shy, but she possessed great courage.

4. She was sometimes called "Lady Lindy" after Charles A. Lindbergh, who made the first solo flight across the Atlantic Ocean in 1927.

5. Always an unconventional dresser, Earhart became famous for the white silk scarf that she often wore with a brown bomber jacket and a pair of trousers.

## Verbs

**Verbs** assert an action or express a condition (see **3**).

*action*
↓
Tall sunflowers *swayed* gracefully.

*condition*
↓
The tall plants *were* sunflowers.

A main verb may have helpers, called **auxiliary verbs,** such as *are, have, will be, do, did.*

> *auxiliary verb*      *main verb*
>         ↘          ↙
> The strong wind *will bend* the sunflowers.

> *auxiliary verbs*     *main verb*
>     ↙ ↘        ↓
> Leif Erickson *may have preceded* Columbus to America.

**Linking verbs** express condition: *appear, become, feel, look, seem, smell, sound, taste.* The most common linking verbs are the various forms of the verb *to be,* such as *is, are, was, were, be, being, been* (see p. 520).

> *linking verb*
>     ↓
> The woman with the plastic fruit on her hat *is* an actress.

> *linking verb*
>     ↓
> In the presence of so many toys, the child *appeared* joyful.

Verbs are either **transitive** or **intransitive** (see pp. 41, 523). (For **verb tenses,** see **4.** For **verbals,** see pp. 21–22, 524).

■ **Exercise 2**

*Identify verbs.*

1. The aroma of fresh popcorn wafted through the building.

2. The flight attendant walked among the passengers and looked for the ring.

3. The symphony was performed by a distinguished orchestra, and the audience was enthusiastic.

4. Hercules performed the twelve labors that Hera demanded.

5. A chance encounter with a celebrity can be an exciting experience..

## Adjectives

**Adjectives** modify a noun or a pronoun by describing, qualifying, or limiting. Most adjectives appear before the word they modify (see **10**).

> *adjective*
> ↓
> *Beautiful* music soothes the soul.

**Predicate adjectives** follow linking verbs and modify the noun or pronoun that is the subject of the sentence.

> *subject*                     *predicate adjective*
> ↓                                   ↓
> The wild blackberries in the fresh pie tasted *delicious*.

The three **articles** *(a, an, the)* are classified as adjectives.

Some **possessive adjectives** have forms that are the same as some **possessive pronouns:** *her, his, its, our, their, your. (Her, our, their,* and *your* have endings with *-s* in the pronoun form.)

**Demonstrative adjectives,** which have exactly the same forms as demonstrative pronouns, are used before the nouns they modify.

*this* dog, *that* dog; *these* dogs, *those* dogs

**Indefinite adjectives** have the same form as indefinite pronouns: for example, *any, each, every, some.*

## Adverbs

**Adverbs** (like adjectives) describe, qualify, or limit other elements in the sentence. They modify verbs (and verbals), adjectives, and other adverbs.

| | |
|---|---|
| **slid** *smoothly* | adverb modifying a verb |
| *very* **fast** race | |
| *undeniably* **true** story | adverbs modifying adjectives |
| ran *very* **fast** | |
| tried *extremely* **hard** | adverbs modifying adverbs |

Sometimes adverbs modify an entire clause or sentence.

*Apparently,* this kind of plastic is stronger than steel.

**Conjunctive adverbs** show the relationship between a sentence or an independent clause and an earlier sentence or clause. Conjunctive adverbs show relationships of addition *(furthermore, moreover),* contrast *(however, nonetheless),* comparison *(similarly),* result *(therefore, thus),* and time *(meanwhile).*

She found a job; *therefore,* she could buy the car.
*However,* she still could not afford to buy the house.

Many adverbs end in *-ly (effectively, curiously),* but not all words that end in *-ly* are adverbs *(lovely, friendly).*

Adverbs often tell how *(slowly, well),* how much *(profusely, somewhat),* how often *(frequently, always),* when *(late, before),* or where *(there, here)*. Adverbs must be identified by their use in the sentence; that is, cannot be memorized.

## ■ Exercise 3

*Identify adverbs and adjectives. Tell what each one modifies. (Remember,* a, an, *and* the *are adjectives.)*

1. Many young couples carefully restore the interiors of old homes that have been almost entirely ignored for many years.

2. The flooring in such homes is often red oak; however, this warm, glowing wood frequently has been covered with dusty old carpeting.

3. To the keen eye of a skilled architect, a gracefully arching cove ceiling is a work of art.

4. Fortunately, most mistakes in decorating can be easily corrected.

5. Nevertheless, owners should be extremely careful before they remove any unique features of a home.

## Conjunctions

**Conjunctions** connect words, phrases, or clauses. They are classified as **coordinating, correlative,** or **subordinating.**

**Coordinating conjunctions**—*and, but, for, nor, or, so, yet*—connect elements that are of equal grammatical rank or importance.

*coordinating conjunction*

The orchestra played selections from *Brahms, Bach,* **and** *Wagner.*

*coordinating conjunction*

*The conductor left the stage,* **and** *he did not return.*

**Correlative conjunctions** are pairs of words that join equal elements (see **18c**). Examples are *both . . . and, either . . . or, not only . . . but also, neither . . . nor.*

*correlative conjunctions*

An athlete needs **not only** *strength* **but also** *discipline.*

**Subordinating conjunctions** introduce a subordinate or dependent clause. Examples are *after, although, as, because, before, even though, if, since, so that, though, unless, until, when, where, while.*

*subordinating conjunction*

**Before** *many painters sell their work,* they labor for years.

Some words (such as *before, after*) may function as conjunctions and also as other parts of speech.

■ **Exercise 4**

*Identify conjunctions and tell whether each is coordinating, correlative, or subordinating.*

1. After the depression had ended, Americans were eager to escape to a glamorous world, and elaborate musical motion pictures provided an escape and a release.

2. When the curtain went up in the theater, the world looked entirely new, so thoughts of unemployment could be left behind.

3. The plots of these films were predictable, but the costumes were elaborate, and the full orchestras were inspiring.

4. Inevitably, the story involved a spirited couple who were separated either by outside forces or by their own stubbornness before they were happily reunited.

5. Both Ginger Rogers and Fred Astaire were accomplished dancers who often starred in these films.

## Prepositions

**Prepositions** connect a noun or pronoun to another word in the sentence and show a relationship, often of time or direction.

the fox *in* the box

(The preposition *in* connects the word *fox* with the word *box*.)

Some of the most common prepositions are *above, across, after, against, along, among, at, before, behind, below, beneath, beside, between, by, from, in, into, of, on, over, through, up, upon, with, without.* Groups of words can also serve as prepositions: *along with, according to, in spite of.*

A preposition introduces a word group called a **prepositional phrase** (see p. 21), which is made up of the preposition, the **object of the preposition,** and **adjectival modifiers** (see p. 20).

> *preposition     object of the preposition*
> ↓          ↓
> The butterfly hovered **above** the brilliant **sunflower.**

> *preposition     object of the preposition*
> ↙ ↘    ↓
> **Along with** the **milk,** she ordered a cheese sandwich.

(See **37h** for a discussion of the idiomatic use of prepositions.)

## ■ Exercise 5

*Identify prepositions, objects of prepositions, and prepositional phrases.*

1. The soprano delivered the aria with supreme effectiveness, a combination of sweetness and sorrow that greatly moved the audience.

2. According to the rules of the game, the contestant at the front of the line must turn quickly and race toward the rear.

3. There on the shelf was the book the hostess had borrowed from her friend.

4. Never in the history of the province had so fortunate an event occurred.

5. After a fast stroll along the deck, the passenger fell into a chair and slept for an hour.

## Interjections

**Interjections** are words that express surprise or strong emotion. They may stand alone or serve as part of a sentence.

> *Ouch!*
> *Ah,* I understand the problem.

Because of their nature, interjections are used more often in speech than in writing, which is generally more deliberate than spontaneous.

## ■ Exercise 6

*Name the part of speech of each word underlined and numbered.*

1   2   3  4  5   6  7  8 9

<u>Oh</u>, <u>people</u> <u>are</u> <u>not</u> <u>always</u> <u>kind</u>, <u>but</u> <u>that</u> <u>is</u> not justification

for bitterness.

$$\overset{10}{\underline{\text{Successful}}}\ \text{athletes}\ \underline{\text{and}}\ \overset{11}{\underset{}{}}\ \overset{12}{\underline{\text{recognizable}}}\ \overset{13}{\underline{\text{politicians}}}\ \text{mingled}$$

```
     10                    11       12          13
Successful athletes and recognizable politicians mingled
 14    15                    16      17          18
with famous writers and prominent actors at the opera,
 19    20
which opened last night.
```

## The Parts of Sentences

A sentence is a group of words that expresses a complete thought. The essential parts of a sentence are a **subject** and a **predicate.**

A **subject** does something, has something done to it, or is identified or described.

| *subject* | *subject* | *subject* |
|---|---|---|
| ↓ | ↓ | ↓ |
| *Birds* sing. | *Songs* are sung. | *Birds* are beautiful. |

A **predicate** expresses what the subject does, what is done to it, or how it is identified or described.

| *predicate* | *predicate* | *predicate* |
|---|---|---|
| ↓ | | |
| Birds *sing*. | Songs *are sung*. | Birds *are beautiful*. |

## Simple subjects, complete subjects, compound subjects

The essential element of a subject is called the **simple subject.** Usually it consists of a single word.

> *simple subject*
> ↓
> The large *balloon* burst.

The subject can be understood rather than actually stated. A director of a chorus might say "Sing," meaning "You sing." But here one spoken word would be a complete sentence.

*understood subject*    *predicate*

[*You*] Sing.

All the words that form a group and function together as the subject of a sentence are called the **complete subject.**

*complete subject*

The large balloon burst.

Subjects can be **compound** (that is, two or more subjects joined by a conjunction) and function together.

*compound subject*

Students and *faculty* cheered.

Of course pronouns as well as nouns may make up compound subjects.

*compound pronoun subject*

*He* and *I* took a stroll.

■ Exercise 7

*Identify the complete subjects. Tell whether each sentence has a simple subject or a compound subject.*

1.  The secretary and the administrative assistant discussed the error on the spreadsheet.

2. The new computer has proven a useful but frustrating tool.

3. Both teachers and students can benefit from using new technology.

4. The soft green screen and a properly adjusted keyboard help the user to feel comfortable.

5. A speedy printer can turn out ten pages every minute.

## Simple predicates, complete predicates, compound predicates

The single verb (or the main verb and its auxiliary verbs) is the **simple predicate**.

*simple predicate*        *simple predicate*

Balloons *soar.*       Balloons *are soaring.*

The simple predicate, its modifiers, and any complements (see pp. 18–20) form a group that is called the **complete predicate**.

*complete predicate*

Balloons *soared over the pasture.*

When two verbs in the predicate express actions or conditions of a subject, they are a **compound predicate**—just as two nouns may be a **compound subject**.

*compound predicate*

Balloons *soared and burst.*

*compound, complete predicate*

Balloons *soared over the pasture and then burst.*

(For errors in predication, see **15b**.)

## ■ Exercise 8

*Identify the verbs (including auxiliaries) and the complete predicates. Tell whether each sentence has a simple predicate or a compound predicate.*

1. The profit margin on the item was unusually slim.

2. All day long the caravan moved through long valleys and over steep hills.

3. Plan for a bright and meaningful future.

4. Playgoers arrived at the theater early and waited eagerly for the box office to open.

5. Children at the picnic played softball, swam in the pool, and rowed on the lake.

## Complements

**Complements** complete the meaning of the sentence and are usually part of the predicate. They are nouns, pronouns, or adjectives. With linking verbs (see p. 7), they function as **predicate**

**adjectives** or **subjective complements.** With transitive verbs, they function as **direct** or **indirect objects** and are always in the objective case (see **9**).

A **predicate adjective** is an adjective that follows a linking verb (see p. 7) and modifies the subject of the sentence, not the verb.

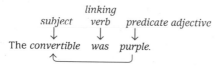

```
                linking
     subject     verb    predicate adjective
        ↓          ↓          ↓
  The convertible  was     purple.
```

A **subjective complement** is a noun that follows a linking verb and renames the subject. (It is also called a **predicate nominative.**)

```
                    subjective complement
                           ↓
  Georgia O'Keefe was a painter.
```

A **direct object** receives the action of a transitive verb. (See p. 48).

```
         verb          direct object
          ↓                 ↓
  The board appointed a new president.
```

An **indirect object** receives the action of the verb indirectly. It comes between the verb and the direct object and tells to whom or for whom something is done.

```
              indirect   direct
     verb       object   object
      ↓           ↓        ↓
  The guest gave the hostess flowers.
```

To identify an indirect object, rearrange the sentence by using the preposition *to* or *for:*

             *direct*    *prepositional*
     *verb*   *object*      *phrase*
      ↓      ↓        ↓
The guest *gave*  *flowers*  *to the hostess.*

(Now *hostess* functions not as the indirect object but as the object of the preposition *to*.)

**Objective complements** accompany direct objects. They modify the object or rename it.

The editor considered the manuscript *publishable*.

The corporation named a former clerk its *president*.

■ **Exercise 9**

*Underline and identify predicate adjectives (PA), predicate nominatives (PN), direct objects (DO), and indirect objects (IO). (Start by finding the verbs and their complements.)*

1. Venice will always be a beautiful place.

2. Those who love art should visit the city.

3. The famous canals are busy and picturesque.

4. Give the gondoliers a good tip and a hearty handshake.

5. The sea vistas are especially stunning, but I was pleasantly surprised when my guide told me to look toward the shore.

## Phrases

A **phrase** is a group of words that does not have both a subject and a predicate. Some important kinds of phrases are **prepositional phrases, verb phrases,** and **verbal phrases.**

### *Prepositional phrases*

**Prepositional phrases** function as adjectives or adverbs.

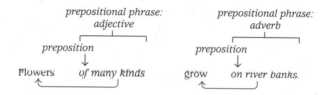

### *Verb phrases*

**Verb phrases** are made up of the main verb and its auxiliaries: *were sitting, shall be going, are broken, may be considered.* Common auxiliary verbs, or "helping" verbs, include *were, shall be, are,* and *may be.*

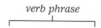

The metaphor *may be considered* an essential tool for writing.

### *Verbal phrases*

**Verbal phrases** are made up of verbals and the words associated with them. A **verbal** is a verb form that does not function as the verb of a clause but rather as a noun, adjective, or adverb. There are three kinds of verbals: **gerunds, participles,** and **infinitives.**

GERUNDS AND GERUND PHRASES

Gerunds always end in *-ing* and function as nouns.

> *gerund phrase as noun*
> *(complete subject)*
> ┌─────┴─────┐
> *gerund*
> ↓
> *Riding a bicycle* is good exercise.

PARTICIPLES AND PARTICIPIAL PHRASES

Participles always function as adjectives. **Present participles** end in *-ing (sailing, watching);* **past participles** usually end in *-ed (hopped, skipped),* but some follow irregular forms *(swum, brought).*

> *participle ending in* -ing *(modifies* she)
> ↓ ┌────────→
> *Swimming steadily*, she reached the shore.
> └────┬────┘
> *participial phrase as adjective*
>
> *participle ending in* -ed *(modifies* bread)
> ↓ ┌────→
> *Overbaked*, the bread was hard.

NOTE: Both gerunds and present participles end in *-ing*, but you can distinguish between them by identifying their function in a sentence. (Gerunds function as nouns; present participles function as adjectives.)

INFINITIVES AND INFINITIVE PHRASES

Infinitives begin with *to,* which is sometimes understood rather than actually stated. They can be used as nouns, adjectives, or adverbs.

USED AS NOUN

        *infinitive phrase*
        *used as subject*

  *infinitive*

*To operate the machine* was simple.

USED AS ADJECTIVE

                *infinitive phrase*
                *modifies* book *(noun)*

        *infinitive* ——

*Charlotte's Web* is a good book *to read to a child.*

USED AS ADVERB

                *infinitive phrase*
                *modifies* reluctant *(adjective)*

        *infinitive*

A true hero is reluctant *to boast of accomplishments.*

*TO* UNDERSTOOD

    Someone must go [*to*] *buy* a tire.

# ■ Exercise 10

*Identify phrases and tell what kind each is—infinitive, gerund, participial, prepositional, or verb.*

1. Taking the census is not as simple as it was two thousand years ago.

2. Candidates for local office will be attending the parent-teacher association meeting to make themselves known and to express their views.

3. Perceiving his opponent's strategy, the world-famous chess player managed to win the game and to do so in record time.

4. Speaking strangely, the young man from the audience said that his will was too strong for him to be hypnotized, but nevertheless he was doing whatever the hypnotist commanded.

5. Insisting that his vegetables were fresh, the gardener refused to lower the price.

## Clauses

A **clause** is a group of words containing a subject and a predicate. Clauses are either **independent** or **dependent (subordinate).**

### *Independent clauses*

An **independent clause** can stand alone grammatically and form a complete sentence. Two or more independent clauses in one sentence can be joined by coordinating conjunctions, conjunctive adverbs, semicolons, or other grammatical devices or punctuation marks.

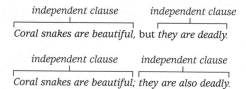

independent clause          independent clause

Coral snakes are beautiful, but *they are deadly.*

independent clause          independent clause

Coral snakes are beautiful; *they are also deadly.*

## Dependent clauses

A **dependent clause** has a subject and a predicate, but it cannot stand alone grammatically because it begins with a subordinating word. Like verbals, dependent clauses can function as nouns, adjectives, and adverbs.

USED AS NOUN (usually subject or object)

> *That the little child could read rapidly* was well known.
> *(noun clause used as subject)*

> The other students knew *that the little child could read rapidly.*
> *(noun clause used as direct object)*

USED AS ADJECTIVE

> Everyone *who completed the race* won a shirt.
> *(modifying pronoun subject* Everyone)

USED AS ADVERB

> *When dusk comes* the landscape seems to hold its breath.
> *(modifying verb* seems)

## The Kinds of Sentences

A **simple sentence** has only one independent clause (but no dependent clause). A simple sentence is not necessarily a short sentence; it may contain several phrases.

> Birds sing.
>
> The bird began to warble a sustained and beautiful song after a long silence.

A **compound sentence** has two or more independent clauses (but no dependent clause).

> *independent clause*        *independent clause*
>
> *Birds sing,* and *bees hum.*

A **complex sentence** has both an independent clause and one or more dependent clauses.

> —————— *dependent clauses* ——————┐ ┌*independent*—
>
> When spring comes and [when] new leaves grow, *migratory birds*
>
> — *clause* ┐
>
> *return north.*

A **compound-complex sentence** has at least two independent clauses and at least one dependent clause. A dependent clause can be part of an independent clause.

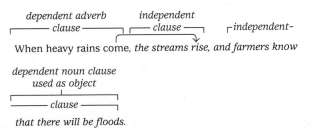

> *dependent adverb*        *independent*
>
> ┌—— *clause* ——┐ ┌—— *clause* ——┐        ┌*independent*-
>
> When heavy rains come, *the streams rise, and farmers know*
>
> *dependent noun clause*
> *used as object*
>
> ┌——— *clause* ———┐
>
> *that there will be floods.*

## ■ Exercise 11

*Underline each clause. Tell whether it is dependent or independent. Identify whether each sentence is simple (S), compound (C), complex (X), or compound-complex (CC).*

1. Arriving at college for the first time can be an unsettling experience.

2. Although I was eager to leave home and be on my own, the prospect of sharing a room with a stranger did not appeal to me.

3. Since I was the oldest child in my family, I had always had the luxury of my own room, and I valued the privacy that my privileged status earned me.

4. I arrived at school, found my way to the dormitory, and met my roommate.

5. When I first met her, I was amazed to discover how much we had in common.

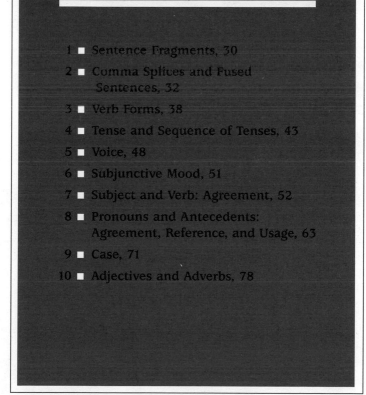

# Sentence Errors

# 1 Sentence Fragments *frag*

Write complete sentences.

Knowledge of what makes up a sentence (and what does not) is basic to an understanding of composition principles (see p. 15). Writing a sentence fragment rather than a complete sentence can therefore be a serious error.

Notice how the fragments below are revised and made into complete sentences.

FRAGMENT (phrase)
Genealogy, the study of family history.

COMPLETE SENTENCE (verb added)
Genealogy is the study of family history.

FRAGMENT (dependent clause)
Although several large rivers have been cleaned up.

COMPLETE SENTENCE (subordinating conjunction *although* omitted)
Several large rivers have been cleaned up.

FRAGMENT (noun and phrase—no main verb)
The green fields humming with sounds of insects.

COMPLETE SENTENCE (modifier *humming* changed to verb *hummed*)
The green fields hummed with sounds of insects.

Fragments are often permissible in dialogue when the meaning is clear.

"See the geese."
"Where?" [fragment]
"Flying north." [fragment]

Fragments are occasionally used for special effects or emphasis.

The long journey down the river was especially pleasant. A time of rest and tranquility.

# ■ Exercise 1

*Write F by fragments, C by complete sentences.*

1. Looking forward to a prosperous and fulfilling career.
2. Increasingly, educators use computers for instructional purposes.
3. Sixty miles from the nearest town.
4. Do not be late.
5. Florida and California, two states that boast of almost constant sunshine.
6. How did people ever do without transparent tape?
7. Not having a care in the world.
8. Travel enlightens.
9. Lightning not striking twice in the same place.
10. "Where the deer and the antelope play."

# ■ Exercise 2

*Underline all fragments in the following passage.*

Students of Franklin D. Roosevelt are agreed that the polio attack of 1921 profoundly changed him. He might have become President without having had to surmount that obstacle, but it is unlikely that he would have been a great President. Or even a good one. Before he was crippled, Roosevelt had been a genial glad-hander. An acceptable politician. Considered lightweight by the pros (men like Al Smith). Too anxious to please. Clumsily ingratiating. His caustic cousin, Alice Roosevelt Longworth, called him a sissy. A mama's boy. He had been sheltered from hardship. Cushioned in privilege. At the least, then, the struggle to walk again—always defeated but never quite given up— toughened Roosevelt. Some say that the suffering deepened his sympathy with others. Who were afflicted.

Adapted from GARRY WILLS
"What Makes A Good Leader?"

# 2 Comma Splices and Fused Sentences  *cs/fus*

Join two independent clauses appropriately, or write two separate sentences.

A **comma splice** or **comma fault** occurs when a comma is used between two independent clauses without a coordinating conjunction.

SPLICE OR FAULT

Human nature is seldom as simple as it appears, hasty judgments are often wrong.

A **fused sentence** or **run-on sentence** occurs when the independent clauses have neither punctuation nor coordinating conjunctions between them.

FUSED OR RUN-ON

Human nature is seldom as simple as it appears hasty judgments are often wrong.

Writers of comma splices and fused sentences fail to indicate the break between independent clauses. Revise in one of the following ways:

1. Use a *period* and write two sentences.

   Human nature is seldom as simple as it appears. Hasty judgments are often wrong.

2. Use a *semicolon* (see also **22**).

   Human nature is seldom as simple as it appears; hasty judgments are often wrong.

NOTE:   Before *conjunctive adverbs* (see **20f** and **22a**), use a *semicolon* to join *independent clauses,* or use a *period* and *begin a new sentence.*

   Production of the item has greatly increased; *therefore,* the cost has come down.

   Production of the item has greatly increased. *Therefore,* the cost has come down.

3. Use a *comma* and a *coordinating conjunction (and, but, for, nor, or, so, yet)* (see also **20a**).

   Human nature is seldom as simple as it appears, *so* hasty judgments are often wrong.

4. Use a *subordinating conjunction* (see pp. 11, 523) and a *dependent clause.*

   **Because** *human nature is seldom as simple as it appears,* hasty judgments are often wrong.

## ■ Exercise 3

*Identify comma splices and fused sentences with* CS *or* FUS. *Correct them. Write* C *by sentences that are already correct.*

1. Russian music flowered late, however, it has developed rapidly during the past two centuries.

2. Few people realize that Sigmund Freud was born in what later became Czechoslovakia, where he spent his early years in poverty.

3. Conflicts almost always exist within a family nevertheless, it is still the most enduring of social units.

4. The bread stuck in the toaster the smoke detector went off the smell of burned toast permeated the apartment.

5. Signs warning about riptides and the undertow were posted on the beach, no one ventured into the water.

## ■ Exercise 4

*Identify each of the following as correct* (C), *a comma splice* (CS), *a fused sentence* (FUS), *or a fragment* (F).

1. The *Savannah* was launched in 1958, it was the first ship to be propelled by nuclear power.
2. Vitamins are necessary for health; however, excessive amounts of some of them are dangerous.

3.  Vitamins are necessary for health. Excessive amounts of some of them, however, dangerous.
4.  Nearly every student of the classical guitar admires the playing of Andrés Segovia, who was essentially self-taught and who gave his first public performance when he was sixteen.
5.  The magnitudes of earthquakes are measured by instruments called seismographs they record movements in the earth's crust.

■ **Exercise 5**

*Identify fragments (F), comma splices (CS), or fused sentences (FUS), and correct them. Write C by correct sentences.*

1.  Some students do not look forward to studying poetry, they probably have not had much experience with it.

2.  Pleasure does seem to make time go faster.

3.  Jane Addams was a famous social reformer she won the Nobel Prize in 1931.

4.  All over the city, new buildings springing up, and the skyline rapidly changing.

5.  Lighthouses are fascinating, many people travel around the country to visit them.

6.  An impressive lighthouse in Florida, tall and stately and painted red.

7. Continual snowfall, increasing the likelihood that classes will be canceled.

8. For many, swimming is more pleasant than jogging, especially in hot weather.

9. The clothing of former times appears strange and comical, that of the present seems appropriate and smart, why is that?

10. Learning to play a musical instrument is difficult when one is young there are often many distractions.

■ Exercise 6

*Follow the instruction for Exercise 5.*

1. Autumn leaves are especially vivid this year. Leaves of bright red and gold.

2. A relatively new kind of surgery is helping many nearsighted people not everyone, though, is a good candidate for the procedure.

3. St. John's is the capital and the port of Newfoundland. An island off the east coast of Canada.

4. Many generous people are naive, they simply do not realize when they are being imposed upon.

5. The Statue of Liberty, a gift from France, is made of copper, it is situated on an island in New York harbor.

6. Microwave ovens are in wide use today.

7. After all, the student argued, any imbecile can use periods and commas, studying punctuation is a complete waste of time.

8. Some families agreed that for two weeks they would keep their television sets turned off the children were surprisingly cooperative.

9. Why anyone would want to go over Niagara Falls in a barrel is puzzling many people have tried it, however.

10. The tortoises on the Galapagos Islands often weigh great amounts. Some as much as five hundred pounds.

# 3 Verb Forms *of*

Use the correct form of the verb.

All verbs have five different forms:

| | |
|---|---|
| BASE FORM (present infinitive) | walk |
| PAST TENSE | walked |
| PAST PARTICIPLE | walked |
| *S*-FORM FOR PRESENT TENSE | walks |
| PRESENT PARTICIPLE | walking |

Most verbs are **regular;** they form the past tense and the past participle by adding *-d* or *-ed* to the base form. Sometimes you will have to double the final consonant of the base verb before adding the *-d* or *-ed*. The dictionary lists only one form, the base (or infinitive) form, when the verb is regular.

| BASE | PAST TENSE | PAST PARTICIPLE |
|---|---|---|
| help | helped | helped |
| smile | smiled | smiled |
| scan | scanned | scanned |
| talk | talked | talked |

Over two hundred verbs in English are **irregular;** they do not follow the standard formula to produce the past tense and past participle forms. For irregular verbs, a dictionary gives four forms: the base (infinitive), past tense, past participle, and present participle.

| BASE | PAST TENSE | PAST PARTICIPLE | PRESENT PARTICIPLE |
|---|---|---|---|
| see | saw | seen | seeing |
| think | thought | thought | thinking |
| freeze | froze | frozen | freezing |

Although you can always use the dictionary to determine the correct form of a verb, memorizing irregular verbs and their forms is more helpful. The following list includes the most commonly used irregular verbs. See also the list of irregular verbs on pp. 510–511.

| BASE | PAST TENSE | PAST PARTICIPLE |
|---|---|---|
| awake | awoke, awaked | awoke, awaked |
| be | was | been |
| begin | began | begun |
| bid (to offer as a price or to make a bid in playing cards) | bid | bid |
| bid (to command, order) | bade, bid | bidden, bid |
| blow | blow | blown |
| break | broke | broken |
| bring | brought | brought |
| build | built | built |
| burst | burst | burst |
| buy | bought | bought |
| choose | chose | chosen |
| come | came | come |
| deal | dealt | dealt |
| dig | dug | dug |
| dive | dived, dove | dived |
| do | did | done |
| drag | dragged | dragged |
| draw | drew | drawn |
| drink | drank | drunk |
| drive | drove | driven |
| drown | drowned | drowned |
| eat | ate | eaten |
| fly | flew | flown |
| freeze | froze | frozen |
| get | got | gotten, got |
| give | gave | given |
| go | went | gone |
| grow | grew | grown |
| know | knew | known |
| lead | led | led |
| lend | lent | lent |

| BASE | PAST TENSE | PAST PARTICIPLE |
|------|-----------|-----------------|
| lose | lost | lost |
| ring | rang | rung |
| run | ran | run |
| see | saw | seen |
| sing | sang | sung |
| sink | sank, sunk | sunk |
| slay | slew | slain |
| sting | stung | stung |
| swim | swam | swum |
| swing | swung | swung |
| take | took | taken |
| teach | taught | taught |
| think | thought | thought |
| throw | threw | thrown |
| wear | wore | worn |
| write | wrote | written |

Some verb forms are especially troublesome. *Lie* is confused with *lay; sit,* with *set;* and *rise,* with *raise.*

*Lie, sit,* and *rise* are intransitive (do not take objects) and have the vowel *i* in the infinitive form and the present tense.

*Lay, set,* and *raise* are transitive (take objects) and have *a, e,* or *ai* as vowels in the infinitive form and the present tense.

| TRANSITIVE | lay (to place) | laid | laid |
|------------|----------------|------|------|
| INTRANSITIVE | lie (to recline) | lay | lain |
| TRANSITIVE | set (to place) | set | set |
| INTRANSITIVE | sit (to be seated) | sat | sat |

In special meanings the verb *set* is intransitive (a hen *sets;* the sun *sets;* and so forth).

| TRANSITIVE | raise (to lift) | raised | raised |
|------------|-----------------|--------|--------|
| INTRANSITIVE | rise (to get up) | rose | risen |

## ■ Exercise 7

*Underline the incorrect verb and write the correct form above it.*
*Write C by correct sentences.*

1. The tourist throwed quarters into the sea, and several islanders dived for them.

2. When Virginia was four, she seen a parade and run along the street to keep up with it.

3. Although the most recently appointed member of the board come in late to the meeting, she done right when she apologized.

4. Before the Bebo family bought a car, they taken few trips; but now that they are use to traveling, they seldom stay home in the summer.

5. The builder lead the owner of the property to the back of the lot and showed him where someone had drug old cars onto his land and had left them there as eyesores.

6. The passenger give a small tip to the cabdriver, who frowned and glared.

7. Abigail sat the potted plant in the window and taken the dog for a walk.

8. The artist laid the brush on the stand after he painted the portrait, and it has been lying there ever since.

9. The carpenter promised to sit the bucket on the tile, but it has been setting on the carpet for a week.

10. The balloon that hanged over the doorway suddenly busted.

## ■ Exercise 8

*Follow the instructions for Exercise 7.*

1. Murray layed around the house all day and then wondered where his time had went.

2. The fall had came and gone, and still he had not wrote his parents.

3. Often he just set in front of his television watching old films that he had saw before.

4. Now and then he eat a candy bar that he had froze in the refrigerator.

5. He been that way for about a month when someone rung his doorbell.

6. It done no good to ring the bell, however, for Murray give no heed.

7. Then, when the visitor just opened the door and walked in, Murray taken one look at him and jumped up in surprise.

8. This were not just any visitor who had busted in but his boyhood hero, his brother, who had flew there from overseas to see him.

9. They talked well into the night, and Murray suddenly seemed to see things much more clearly than he had.

10. In a few days, Murray's terrible spell was broke, and he begun to put his life back together.

# 4 Tense and Sequence of Tenses    *t/shift*

Use appropriate forms of verbs to express time sequences. Avoid incorrect shifts in tense.

Verbs use tense to express the time of an action or condition.

**4a**  Use the **present tense** to express an action or condition that is currently occurring, regularly occurring, or consistently true. The present tense may also be used to refer to events occurring in a literary work or other works of art.

> The student *walks* to the library. [currently occurring]
> Every Saturday he *does* his laundry. [regularly occurring]

In 1851, Foucault proved that the earth *rotates* on its axis. [consistently true]

In Kate Chopin's *The Awakening,* Edna *decides* to leave her husband in an attempt to find her freedom. [events in a work of literature]

**4b**   Use the **past tense** to express an action that was completed or a condition that occurred before the time of writing.

The student *walked* to the library yesterday. [completed]

Ancient Greeks *believed* that the earth *was* motionless. [once believed, but now disproved]

**4c**   Use the **future tense** to express an action or condition expected to occur after the time of writing.

The most common future tense form combines *will* with the base form of the verb.

The student *will walk* to the library tomorrow.

**4d**   Use **progressive tenses** to show that an action or condition is ongoing.

The three progressive tenses are **present progressive, past progressive,** and **future progressive.** Progressive forms use the present participle form of the verb (the *-ing* form) with auxiliary verbs. (See p. 7.)

PRESENT PROGRESSIVE

The computer *is sending* an error message. [event taking place currently]

PAST PROGRESSIVE

> The computer *was sending* an error message all last week. [event ongoing during a certain time frame in the past]

FUTURE PROGRESSIVE

> Because more professors *are starting* to use computers, students *will be expecting* to receive their grades on E-mail. [future event that will be ongoing and that depends on another action]

**4e** Use **perfect tenses** to indicate one time or action completed before another.

The three perfect tenses are **present perfect, past perfect, and future perfect.**

PRESENT PERFECT WITH PRESENT

> I *have paid* the rent; therefore, I **am moving** in.

The controlling time word sometimes is not a verb.

> I *have paid* the rent **already.**

PAST PERFECT WITH PAST

> I *had paid* the rent; therefore, I **moved** in.
> I *had bought* my ticket before the bus **came.**

CAUTION:   Do not use the past perfect tense when the past tense will suffice.

NOT

> I *had ordered* shrimp, but the waiter brought me scallops.

BUT

> I *ordered shrimp,* but the waiter brought me scallops.

FUTURE PERFECT WITH FUTURE

> I *shall have eaten* by the time we **go.** [The controlling time word, *go,* is
> present tense in form but future in meaning.]

> I *shall have eaten* by **one o'clock,** [Note that the controlling time words
> do not include a verb.]

The future perfect is rare. Usually the simple future tense is used
with an adverb phrase or clause.

RARE

> I shall have eaten before you go.

MORE COMMON

> I shall eat before you go.

NOTE:   The perfect participle also expresses an action that precedes another action.

> *Having laughed* at length, the audience finally quieted down.

**4f**   Use the **present infinitive** (*to* and the base form of a
verb) when it expresses action that occurs at the same time
as that of the controlling verb.

NOT

> I wanted *to have gone.*

BUT

> I wanted *to go.*

NOT

> I had expected *to have met* my friends at the game.

BUT

> I had expected *to meet* my friends at the game.

NOT

> I would have preferred *to have waited* until they came.

BUT

> I would have preferred *to wait* until they came.


**4g**  Be consistent in the use of tenses. Do not shift tenses within a sentence or a paragraph without good cause.

TWO PAST ACTIONS

> The furniture maker *sanded* each piece with thorough care and only then *applied* [*not applies*] the stain.
>
> The child *smiled* broadly when she *saw* the clown approaching her.

TWO PRESENT ACTIONS

> As one *makes* new friends, school life *becomes* interesting.


■ **Exercise 9**

*Underline incorrect verbs or verbals, and write the corrections above them. Write C by correct sentences.*

1. The judge listened to attorneys for both sides and then makes his ruling.

2. North American Indians are said to have originated the game of lacrosse.

3. In novels by Charles Dickens, a young man sometimes discovered that his birthright is far more advantageous than he had imagined.

4. Having composed a rough draft on his computer, the author sat back in his chair and stares at the ceiling.

5. The horses rushed out of the starting gate and head for the first turn.

6. In looking back, public officials almost always say they would have preferred to have remained private citizens.

7. The soprano met with the civic group, agreed to sing a solo at the Christmas concert, and begins to search for suitable music.

8. The periodic table shows that the symbol for the element mercury was Hg, not Me.

9. At one time, many people thought that the world is flat.

10. In Gérôme's painting *The Cadet,* the young man had a slight sneer on his face.

# 5 Voice *vo*

Use the active voice for conciseness and emphasis.

A transitive verb (see p. 40) is in either the active or the passive voice. (An intransitive verb does not have voice.) When the subject

acts, the verb is active. In most sentences the actor is more important than the receiver.

PASSIVE        A positive impression *was made* on the employees by
               the new chief executive officer.

USE ACTIVE     The new chief executive officer *made* a positive
               impression on the employees.

PASSIVE        A brief but highly effective speech *was delivered* by
               the valedictorian.

USE ACTIVE     The valedictorian *delivered* a brief but highly effec-
               tive speech.

In the sentences above, the active voice creates a more vigorous style.
    When the subject is acted upon, the verb is passive. A passive
verb is useful when the performer of an action is unknown or
unimportant.

    The book about herbs *was misplaced* among books about cosmetics.

The passive voice can also be effective when the emphasis is on the
receiver, the verb, or even a modifier.

    The police *were* totally *misled*.

In most situations, however, active verbs are more forceful and
concise.

■ Exercise 10

*Rewrite the following sentences. Change from passive to active voice.*

1. The tree was quickly climbed by the white-handed gibbon,

   and the lions below were glared at.

2. A television program was watched by the three children as cake was brought to them by their mother.

3. The road had been traveled many times by the reporter, but the old house had never before been noticed by her.

4. A beautiful sunrise was witnessed by the shipwrecked sailors, and at that very moment a ship dispatched to rescue them was spotted by them.

5. The cameras that were used by the anthropologists to take pictures of the ancient village were destroyed by the leaders of the tribe.

■ **Exercise 11**

*Change the voice of the verb when it is ineffective. Rewrite the sentence if necessary. Write E by sentences in which the verb is effective.*

1. The appearance of the yard was dramatically improved by mowing the grass and pruning the hedges.

2. Both passengers were thrown clear, and they walked away uninjured.

3. The horse lost the race because the shoe was improperly nailed to the hoof.

4. The students were told by their scuba-diving instructor that a good time could be had by all if they simply took proper precautions.

5. Some people are made hungry by cooking shows on television.

# 6 Subjunctive Mood   *mo*

Use the subjunctive mood to express wishes, orders, and conditions contrary to fact (see **Mood**, p. 520).

WISHES

    I wish that tomorrow *were* here.

ORDERS

    The instructions are that ten sentences *be* revised.

CONDITIONS CONTRARY TO FACT

    If I *were* a little child, I would have no responsibilities.

    If I *were* you, I would not go.

    *Had* the weather *been* good, we would have gone to the top of the mountain.

In modern English the subjunctive survives mainly as a custom in some expressions.

SUBJUNCTIVE

The new manager requested that ten apartments *be* remodeled.

SUBJUNCTIVE NOT USED

The new manager decided to have ten apartments remodeled.

## ■ Exercise 12

*Change verbs to the subjunctive mood when appropriate.*

1. This house would sell at a higher price if it was not painted purple.

2. Let us stipulate in our will that our children are well cared for.

3. If it was not for the ozone layer, ultraviolet rays would be more destructive.

4. I demand that my money is refunded!

5. Many people wish that the entire world was at peace.

# 7 Subject and Verb: Agreement    *agr*

Make subjects and verbs agree.

Subjects and their verbs must agree in number; that is, a singular subject takes a singular verb, and a plural subject takes a plural verb.

**7a**   Use a singular verb with a singular subject.

The *-s* or *-es* ending of the present tense of a verb in the third person *(she talks, he wishes)* indicates the singular.

> *singular*
> ↓   ↓
> The *door* **opens.**

> *singular*
> ↓   ↓
> The *noise* **disturbs** the sleepers.

**7b**   Use a plural verb with a plural subject.

> *plural*
> ↓   ↓
> The *doors* **open.**

> *plural*
> ↓   ↓
> The *noises* **disturb** the sleepers.

**7c**   Use a plural verb ordinarily with a compound subject (see p. 16).

> *Work and play* **are** not equally rewarding.
>
> *Golf and polo* **are** usually outdoor sports.

EXCEPTION:   Compound subjects connected by *and* but expressing a singular idea take a singular verb.

> *The rise and fall* of waves **draws** a sailor back to the sea.
> When the children are in bed, *the tumult and shouting* **dies.**
> *The coach and history teacher* **is** Mr. Silvo.

**7d**    After a compound subject with *or, nor, either . . . or, neither . . . nor, not . . . but, not only . . . but also,* make the verb agree in number and person with the nearer part of the subject (see **8d**).

NUMBER

Neither the *photographs* nor the *camera* **was** damaged by the fire.

Either *fans* or an *air conditioner* **is** necessary.

Either an *air conditioner* or *fans* **are** necessary.

PERSON

Neither *you* nor your *successor* **is** affected by the new regulation.

**7e**    Do not allow phrases or clauses that come between a subject and a verb to affect the number of a verb.

Connectives like *along with* and *as well as* are not coordinating conjunctions but prepositions that take objects; they do not form compound subjects. Other such words and phrases include *in addition to, including, plus, together with, with.*

SINGULAR SUBJECT, INTERVENING PHRASE, SINGULAR VERB

The *pilot* as well as all his passengers *was* rescued.

Written with a coordinating conjunction, the sentence takes a plural verb.

The *pilot* **and** his *passengers* *were* rescued.

NOTE:    Do not be confused by inversion.

From kind acts **grows** [not *grow*] *friendship.*

**7f**  Use a singular verb with a collective noun that refers to a group as a unit; use a plural verb when the members of a group are thought of individually.

A collective noun names a class or group: *congregation, family, flock, jury.* When the group is regarded as a unit, use the singular.

The *audience* at a concert sometimes **determines** the length of the performance.

When the group is regarded as separate individuals, use the plural.

The *audience* at a concert **vary** in their reactions to the music.

**7g**  Use a singular verb with most nouns that are plural in form but singular in meaning.

*Economics* and *news* (and other words like *genetics, linguistics,* etc.) are considered singular.

*Economics* **is** often thought of as a science.

The *news* of a cure **is** encouraging.

*Scissors* and *trousers* are treated as plural except when used after *pair.*

The *scissors* **are** dull.

That *pair* of scissors **is** dull.

The *trousers* **are** pressed and ready to wear.

An old *pair* of trousers **is** sometimes stylish.

Other nouns that cause problems are *athletics, measles,* and *politics.* When in doubt, consult a dictionary.

**7h**   Use singular verbs with indefinite pronouns *(anybody, anyone, each, either, everybody, everyone, neither, no one, nobody, one, somebody, someone).*

*Neither* of the explanations **was** satisfactory.

*Everybody* **has** trouble choosing a subject for a paper.

*Each* of the students **has** chosen a subject.

**7i**   Use a singular or a plural verb with words such as *all, some, part, half* (and other fractions), depending on the noun or pronoun that follows.

*singular*

*Some* of the *sugar* **was** spilled on the floor.

*plural*

*Some* of the *apples* were spilled on the floor.

*singular*

*Half* of the *money* **is** yours.

*plural*

*Half* of the *students* **are** looking out the window.

*The number* is usually singular.

*singular*

The *number* of *translations* **was** never determined.

*A number* when used to mean *some* is always plural.

*plural*

A *number* of translations of the phrase **were** suggested.

**7j**  In sentences beginning with *there* or *here* followed by verb and subject, use a singular or a plural verb, depending on the subject.

*There* and *here* (never subjects of a sentence) are sometimes **expletives** used when the subject follows the verb.

*verb*        *subject*

There **was** a long *interval* between the two discoveries.

There **were** thirteen *blackbirds* perched on the fence.

Here **is** a *principle* to remember.

Here **are** two *principles* to remember.

The singular *is* may introduce a compound subject when the first noun is singular after an expletive.

There **is** a *swing and a footbridge* in the garden.

NOTE:   In sentences beginning with *It,* the verb is always singular.

*It* **was** many years ago.

**7k** Make a verb agree with its subject, not with a subjective complement.

*NO*

*His horse* and *his dog* **are** his main source of pleasure.

*NO*

His main *source* of pleasure **is** his horse and his dog.

**7L** After a relative pronoun *(who, which, that),* use a verb with the same person and number as the antecedent.

*antecedent*      *relative pronoun* →    **verb of relative pronoun**

*Those who* → **were** invited came.

*We who* → **are** about to die salute you.

The *costumes that* → **were** worn in the ballet were dazzling.

He was the *candidate who* → **was** able to carry out his pledges.

He was one of the *candidates who* → **were** able to carry out their pledges.

BUT

He was the only *one* of the candidates *who* → **was** able to carry out his pledges.

**7m** Use singular verbs with titles and with words used as words.

*Pride and Prejudice* **deserves** a close reading.

"Hints from Heloise" **is** a syndicated newspaper column.

*Oodles* **is** an informal word.

**7n** Use singular verbs with expressions of time, money, measurement, weight, and volume and also with fractions *(half, quarter, part)* when the amount is considered a unit.

*Two tons* **is** a heavy load for a small truck.

*Forty-eight hours* **is** a long time to go without sleep.

■ **Exercise 13**

*Correct any verb that does not agree with its subject. Write C by any correct sentence.*

1. A ship and an airplane affords different ways of crossing the Atlantic Ocean.

2. "Jingle Bells" have given many children pleasure at Christmastime.

3. The screech of seagulls mingle with the sound of the waves.

4. Several different careers are now open to those who major in the humanities.

5. A book on statistical methods in the social sciences together with several essays are required reading in the course.

6. Molasses were used in a great number of early New England recipes.

7. Childish sentences or dull writing are not improved by a sprinkling of dashes.

8. Neither of the senator's two daughters want a career in politics, which have always attracted other members of the family.

9. Either cash or credit card are acceptable when paying for the tickets to the concert.

10. The breaking of laws bring penalties.

## ■ Exercise 14

*Correct any verb that does not agree with its subject. Mark C by any correct sentence.*

1. Garrison Keillor's *Lake Wobegon Days* recount the history of a mythical Minnesota town.

2. The stories from this humorous book captures small-town American life.

3. A number of the characters have been developed from Keillor's PBS radio program, "A Prairie Home Companion."

4. In the printed version of the stories, Keillor's warm voice and infectious laugh is missing, his infrequent verbal slips do not occur, and the sound effects are missing.

5. Dorothy of the Chatterbox Cafe together with Ralph of Ralph's Pretty Good Grocery appear in the book.

6. Neither Dorothy nor Ralph serve customers with modern efficiency.

7. One of the daily specials at the Chatterbox often is tuna hotdish, the odor of which drift across Main Street.

8. Neither the codfish nor the pork loins at Ralph's is very fresh.

9. At Christmastime, smelly tubs of dried cod soaked in lye and called *lutefisk* appears at Ralph's meat counter.

10. Politics were always less important in Lake Wobegon than sports and religion were.

■ **Exercise 15**

*Correct any verb that does not agree with its subject.*

1. The novel *Glittering Images* deal with leaders in the Anglican Church in the 1930's.

2. The difference between illusion and reality play a major role in the story.

3. The title refers to illusions that has to be abandoned as reality intrude.

4. Susan Howatch is one of those novelists who has been trained in law.

5. Some of her characters (the exact number of them are not important) is based on real people.

6. The author, along with her husband, live in Surrey, England.

7. Each of the pages were carefully written and revised.

8. Neither superficiality nor verbosity damage her writing.

9. Great technical skill and philosophical depth is rare in a popular novelist.

10. Four hundred pages are a substantial amount to read, but the book move along swiftly.

# 8 Pronouns and Antecedents: Agreement, Reference, and Usage   *agr/ref*

Make pronouns agree with their antecedents and refer to those antecedents clearly.

A singular pronoun refers to a singular antecedent (see p. 514); a plural pronoun refers to a plural antecedent. A pronoun should refer clearly to a definite antecedent.

**8a**  Use a singular pronoun with a singular antecedent.

The *debater* made **her** point eloquently.

*Water* finds **its** own level.

**8b**  Use a plural pronoun with a plural antecedent.

The *debaters* made **their** points eloquently.

Three *rivers* spread beyond **their** banks.

**8c**  In general, use a plural pronoun to refer to a compound antecedent linked with *and*.

The *owner* and the *captain* refused to leave **their** distressed ship.

If two nouns designate the same person, the pronoun is singular.

The *owner and captain* refused to leave **his** distressed ship.

**8d**  After a compound antecedent linked with *or, nor, either . . . or, neither . . . nor, not . . . but, not only . . . but also,* make a pronoun agree with the nearer part of the antecedent (see **7d**).

> Neither the chess *pieces* nor the *board* had been placed in **its** proper position.
> Neither the *board* nor the chess *pieces* had been placed in **their** proper positions.

A sentence like this written with *and* is less artificial and stilted.

> The chess *pieces* **and** the *board* had not been placed in **their** proper positions.

**8e**  Use a singular pronoun with a collective noun antecedent when the members of the group are considered as a unit, but use a plural pronoun when they are thought of individually.

A UNIT
> The *committee* presented *its* report.

INDIVIDUALS
> The *committee,* some of *them* smiling, filed into the room and took *their* seats.

**8f**  Use singular pronouns with such singular antecedents as *each, either, neither, one, no one, everyone, someone, anyone, nobody, everybody, somebody,* and *anybody.*

> Not *one* of the linemen felt that **he** had played well.

Be consistent in number within the same sentence.

INCONSISTENT

> *singular*      *plural*
>   ↓          ↓
> *Everyone* takes **their** seats.

CONSISTENT

> *Everyone* who is a mother sometimes wonders how **she** will survive
> the day.

Traditionally the pronouns *he* and *his* were used to refer to both
men and women when the antecedent was unknown or representative of both sexes: "Each person has to face **his** own destiny." *He*
was considered generic; that is, a common gender. Today this usage
has changed because many feel that it ignores the presence and
importance of women: there are as many "she's" as "he's" in the
world, and one pronoun should not be selected to represent both.
Writers should be sensitive to this issue. Following are three alternatives to using the masculine pronoun:

1.  Make the sentence plural.

    All *persons* have to face *their* own destinies.

2.  Use *he or she* (or *his or her*).

    Each person has to face *his or her* own destiny.

3.  Use *the* or avoid the singular pronoun altogether.

    Each person must face *the* future.
    Each person must face destiny.

**8g**  Make pronouns refer clearly to definite antecedents.

Pronouns should not refer vaguely to an entire sentence or to a
clause or to unidentified people.

VAGUE

> Some people worry about wakefulness but actually need little sleep. *This* is one reason why they have so much trouble sleeping.

*This* could refer to the worry, to the need for little sleep, or to psychological problems or other traits that have not even been mentioned.

CLEAR

> Some people have trouble sleeping because they lie awake and worry about their inability to sleep.

*It, them, they,* and *you* are sometimes used as vague references to people and conditions that need more precise identification.

VAGUE

> *They* always get *you* in the end.

The problem here is that the pronouns *they* and *you* and the sentence are so vague that the writer could mean almost anything pessimistic. The sentence could refer to teachers, deans, government officials, or even all of life.

NOTE: At times writers let *this, which,* or *it* refer to the whole idea of an earlier clause or phrase when no misunderstanding is likely.

> The grumbler heard that his boss had called him incompetent. *This* made him resign.

Make sure that a pronoun refers clearly to one antecedent, not uncertainly to two.

UNCERTAIN

> The agent visited her client before *she* went to the party.

CLEAR

Before the client went to the party, the agent visited her.

■ **Exercise 16**

*Revise sentences in which pronouns do not agree with their antecedents. Write C by any correct sentence.*

1. At the roadblock, each motorist had to show their driver's license.

2. The jury reached their verdict after four hours.

3. The drifter, along with his irresponsible relatives, never paid back a cent they borrowed.

4. Neither my mother nor my aunt could start their car on that cold morning.

5. Candy and cookies are not known for its nutritional value.

6. Search the field for either a daisy or a dandelion so that we can observe it.

7. Everyone must pay their income taxes.

8. The League of Nations failed because they never received full support from member countries.

9. None of the bridesmaids believed that they would have occasion to wear their lavender dress again.

10. In the early days of the West, almost every man could ride their horse skillfully.

■ **Exercise 17**

*Revise sentences in which pronouns do not refer clearly to one definite antecedent.*

1. On the night of July 14, the patriots stormed the doors of the prison, and they were smashed.

2. There was a rumor going around that worms constituted part of the hamburger meat, and it was nasty.

3. Cattle egrets are so named because they are frequently seen with cows, but they do not seem to mind them.

4. It says in today's paper that a cure is being sought for laziness.

5. They tell us that we must pay taxes whether we want to or not.

6. The poet is famous, but it does not put food on the table.

7. The professor did not push his point further with the student because he was embarrassed.

8. According to dealers in rare coins, they do not put as much silver in coins as they used to.

9. Some photographers have a talent for getting small children to sit quietly; still, they can be impatient.

10. Garth's little sister threw his CD player against the mirror with such force that it broke.

**8h** Use *which* to refer to animals and things, *who* and *whom* to refer to persons and sometimes to animals or things called by name, and *that* to refer to animals or things and sometimes to persons.

The *man* **who** was taking photographs is my uncle.

The *dog,* **which** sat beside him, looked happy.

*Secretariat,* **who** won the Kentucky Derby, will be remembered as one of the most beautiful horses of all time.

Sometimes *that* and *who* are interchangeable.

A *mechanic that (who)* does good work stays busy.

A *person that (who)* giggles is often revealing embarrassment.

NOTE:    *Whose* (the possessive form of *who*) may be less awkward than *of which,* even in referring to animals and things.

The *car* **whose** right front tire blew out came to a stop.

**8i**  Use pronouns ending in *-self* or *-selves* only in sentences that contain antecedents for the pronoun.

CORRECT INTENSIVE PRONOUN

The cook *himself* washed the dishes.

FAULTY

The antiques dealer sold the chair to my roommate and *myself*.

CORRECT

The antiques dealer sold the chair to my roommate and *me*.


■ **Exercise 18**

*Revise sentences that contain errors in usage of pronouns.*

1.  The early navigator which discovered the island thought that he had found a new continent.

2.  Was the first section of the essay written by your coauthor or by yourself?

3.  I bought tickets for my wife and myself.

4.  That woman in the red dress is the one which inherited a fortune and gave it away.

5.  The green frog uses his tongue to catch insects.

# 9 Case   *c*

Use correct case forms.

Case expresses the relationship of pronouns *(I, me)* and nouns to other words in the sentence by the use of different forms. The case of a noun or a pronoun is determined by the function of that word in a sentence. Pronouns serving as **subjects** or **subjective complements** (see p. 19) are in the **subjective case**. Those serving as **objects** are in the **objective case**. Those that show **possession** are in the **possessive case**. Nouns are changed in form only for the possessive case *(child, child's)*.

Following is a chart of the cases of pronouns:

PERSONAL PRONOUNS

| *Singular* | *Subjective* | *Possessive* | *Objective* |
|---|---|---|---|
| First person | I | my, mine | me |
| Second person | you | your, yours | you |
| Third person | he, she, it | his, her, hers, its | him, her, it |

| *Plural* | *Subjective* | *Possessive* | *Objective* |
|---|---|---|---|
| First person | we | our, ours | us |
| Second person | you | your, yours | you |
| Third person | they | their, theirs | them |

RELATIVE OR INTERROGATIVE PRONOUNS

| *Number* | *Subjective* | *Possessive* | *Objective* |
|---|---|---|---|
| Singular | who | whose | whom |
| Plural | who | whose | whom |

To determine case, decide how a word is used in its own clause—for example, whether it is a subject or a subjective complement or an object.

**9a**   Use the subjective case for subjects and for subjective complements that follow linking verbs.

SUBJECT

> After seven years, my former *roommate* and *I* [not *me*] had much to talk about.
> It looked as if my *friend* and *I* [not *me*] were going to be roommates again.

SUBJECTIVE COMPLEMENT

> The fortunate ones were *you* and *I.*

In speech, *it's me, it's us, it's him,* and *it's her* are sometimes used. These forms are not appropriate for formal writing.

**9b**   Use the objective case for a direct object, an indirect object, or the object of a preposition.

DIRECT OBJECT

> The magician *amazed* Yolanda and **me** [not I].

INDIRECT OBJECT

> Mozart *gave* **us** great music.

OBJECT OF PREPOSITION

> The director had to choose *between* **her** and **me** [not **she** and I].

NOTE:   Be careful about the case of *we* or *us* before a noun.

FAULTY

> A few *of* **we campers** learned to cook.

CORRECT

> A few *of* **us campers** learned to cook.

When in doubt, test by dropping the noun:

A few *of* **us** learned to cook.

CORRECT (when pronoun is subject)
*We* campers learned to look.

## 9c Use the objective case for subjects and objects of infinitives.

*subject of infinitive*
The editors considered **her** *to be* the best reporter on the staff.

## 9d Use the same case for an appositive and the word to which it refers (see 23c).

The case of a pronoun appositive depends on the case of the word it refers to.

SUBJECTIVE
*Two delegates*—Esther Giner and I—were appointed by the president.

OBJECTIVE
The president appointed two *delegates*—Esther Giner and **me**.

## 9e After *than* or *as* in an elliptical (incomplete) clause use the same case as if the clause were completely expressed.

*subject of understood verb*
No one else in the play was as versatile as **she** (*was*).

object of understood subject and verb
↓
Her fellow actors respected no one more than *(they respected)* **her.**

## 9f  Use *who* and *whom* correctly.

Although in speech the distinction between *who* and *whom* is not made as often as formerly, it is still important in some circles. In educated writing, using *who* and *whom* in the proper case is essential. *Who* is used for the subjective case; *whom,* for the objective case.

IN QUESTIONS

*subjective case*

subject  verb  object
↓       ↓     ↓
*Who* conceived the idea of a committee?

When words intervene between the pronoun and the main verb, determining the case can be more difficult.

[*Who* or *Whom*?] do they say conceived the idea of a committee?

Mentally cancel the intervening words.

*Who* ~~do they say~~ conceived the idea of a committee?

*objective case*

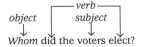

object              subject
↓                    ↓
*Whom* did the voters elect?

To choose the correct pronoun in a dependent clause, first identify the clause; then determine the function of the pronoun **within that** clause: if it serves as subject or subjective complement within the clause, use *who* (or *whoever*); if it serves as an object, use *whom* (or *whomever*). Do not confuse the function of the relative pronoun in its clause with the function of the clause as a whole in the sentence.

IN DEPENDENT CLAUSES

**subjective case**

> *dependent clause*
> subject
> ↓
> I saw who started the fire.

> *dependent clause*
> subject
> ↓
> The car is there for whoever will use it.

**objective case**

> *dependent clause*
> object subject verb
> ↓   ↓   ↙ ↘
> Make an appointment with your English professor, whom you will like.

> *dependent clause*
> object   subject   verb
> ↓     ↓    ↓
> For whomever the bell tolls, it tolls softly.

**9g** Indicate the possessive case by using an *of* phrase or an apostrophe (see 33).

The motto *of the club* was written in Latin.

The club*'s* motto was written in Latin.

**9h** Use the possessive case for pronouns and nouns preceding a gerund.

*gerund*
↓
None of my friends had heard about **my** [not **me**] *leaving*.

*gerund*
↓
The baby's *crying* could be heard throughout the building.

BUT

A noun before a gerund (see p. 22) need not be possessive when the noun is **plural**:

There is no rule against **people** *working* overtime.

when the noun is **abstract**:

I object to **emotion** *overruling* judgment.

when the noun denotes an **inanimate object**:

The police officer did not like the **car** *being* parked in the street.

When a verbal is a participle (see p. 22) and not a gerund, a noun or pronoun preceding it is in the objective case. The verbal functions as an adjective.

No one saw **anyone** *running* from the scene.

■ Exercise 19

*Underline the correct word.*

1. Serve a free meal to (whoever, whomever) is hungry.

2. Margo did not tell her mother (who, whom) she had invited to dinner.

3. Will the delegate from the Virgin Islands please indicate (who, whom) is to receive her delegation's votes?

4. (Who, Whom) do historians believe invented ink?

5. The Muses were supposed to be the nine daughters of Zeus, each of (who, whom) reigned over a different art or science.

6. (Whoever, Whomever) is elected will have to deal with the problem.

7. There was in those days in Paris a singer (who, whom) the secret police knew was a double agent.

8. On the platform stood the man (who, whom) they all believed had practiced witchcraft.

9. On the platform stood the man (who, whom) they all accused of practicing witchcraft.

10. The speaker defended his right to talk critically of (whoever, whomever) he pleased.

■ **Exercise 20**

*Cross out the incorrect forms of pronouns and nouns, and write in the correct forms.*

1. No one was a better sales manager than her.

2. Her's was a rare talent for dealing with people.

3. She was always kind to my mother and I.

4. Just between you and I, whom is it that the judges are favoring?

5. Evan and him appeared just before sunset in that old car, waving and shouting at Eileen and me.

6. The physician said that he had not objected to the employee returning to work.

7. It was plain to we students that the professor was delivering the same lecture that we had heard the week before.

8. If the ordinance were passed granting a right of way to you and I, whose responsibility would it be to make sure that we are allowed access?

9. Between she and I there never were any secrets; at least, that is what she told me.

10. I decided to allow nothing to interfere with me studying.

# 10 Adjectives and Adverbs *adj/adv*

Use adjectives to modify nouns and pronouns; adverbs to modify verbs, adjectives, and other adverbs.

Most adverbs end in *-ly*. Only a few adjectives (such as *lovely, holy, manly, friendly*) have this ending. Some adverbs have two forms, one with *-ly* and one without: *slow* and *slowly, loud* and *loudly.* Most adverbs are formed by adding *-ly* to adjectives: *warm, warmly; pretty, prettily.*

**10a** Use an adverb, not an adjective, to modify a verb, an adjective, or another adverb.

Choosing correctly between adverbs and adjectives in some sentences is simple.

> They stood *close.* [adverb, modifies verb *stood*]
> She was a *close* relative. [adjective, modifies noun *relative*]

Adjectives do not modify verbs, adverbs, or other adjectives. Distinguish between *sure* (adjective) and *surely* (adverb); *easy* and *easily; good* and *well; real* and *really.*

**surely**
You ~~sure~~ cannot count on an alarm clock.

        **easily**
I could have passed the examination ~~easy.~~

  **well**
Allen did ~~good~~ on his final examination.

    **really**
The fireworks were ~~real~~ impressive.

**10b** Use a predicate adjective, not an adverb, after a linking verb (see p. 8) such as *appear, be, become, feel, look, seem, smell, sound, taste.*

ADJECTIVE

*He* feels **bad**. [He is ill or depressed. An adjective modifies a pronoun.]

ADVERB

He *reads* **badly**. [*Reads* is not a linking verb. An adverb modifies a verb.]

ADJECTIVE

The *tea* tasted **sweet**. [*Sweet* describes the tea.]

ADVERB

She *tasted* the tea **daintily**. [*Daintily* tells how she tasted the tea.]

**10c** Use an adjective, not an adverb, to follow a verb and its object when the modifier refers to the object, not to the verb.

Verbs like *keep, build, hold, dig, make, think* are followed by a direct object and a modifier. After verbs of this kind, choose the adjective or the adverb form carefully.

ADJECTIVES—MODIFY OBJECTS

Keep your *clothes* **neat**.
Make the *line* **straight**.

ADVERBS—MODIFY VERBS

*Arrange* your clothes **neatly** in the closet.

*Draw* the line **carefully**.

**10d**  Use the comparative to refer to two things, the superlative to refer to more than two.

The comparative degree of most short modifiers is formed by adding *-er;* the superlative, by adding *-est.*

Both cars are fast, but the small one is *faster.* [comparative degree—one object compared to or contrasted with one other]

All three cars are fast, but the small one is the *fastest.* [superlative degree—one object compared to or contrasted with more than one other.]

Use *more* and *most* (or *less* and *least*) before long modifiers.

|  | COMPARATIVE | SUPERLATIVE |
|---|---|---|
| ADJECTIVES |  |  |
|  | *-er/-est* | |
| dear | dearer | dearest |
| pretty | prettier | prettiest |
|  | *more/most* | |
| pitiful | more pitiful | most pitiful |
| grasping | more grasping | most grasping |
|  | *less/least* | |
| expensive | less expensive | least expensive |
| ADVERBS |  |  |
|  | *-er/-est* | |
| slow | slower | slowest |

ADVERBS

*more/most*

rapidly            more rapidly            most rapidly

*less/least*

sensitively        less sensitively        least sensitively

Some adjectives and adverbs have irregular forms: *good, better, best; well, better, best; little, less, least; bad, worse, worst.* Consult a dictionary.

## 10e   Avoid double comparatives and superlatives.

Using both *-er* and *more,* or *-est* and *most,* creates illogicalities called **double comparatives** and **double superlatives**.

> **earlier**
> Five o'clock is ~~more earlier~~ than most people wish to get up.

> **earliest**
> Three o'clock is the ~~most earliest~~ that I have ever gotten up.

## 10f   Do not apply comparative and superlative degrees to absolute concepts.

Some adjectives and adverbs describe **absolute concepts**. Death, for example, is absolute; one thing cannot be *deader* than another or the *deadest* of a group.

ILLOGICAL
> Denise is the *most unique* person of my acquaintance.

*Unique* is an absolute concept, meaning "one of a kind":

> Among my acquaintances, Denise is *unique.*

ILLOGICAL

> The smaller diamond is *more perfect* than the larger one.

ACCEPTABLE

> The smaller diamond is *more nearly perfect* than the larger one.

# 10g  Avoid double negatives.

**Double negatives** actually mean the opposite of what is intended. One negative is sufficient; two, unacceptable.

NOT

> She could *not hardly* keep from screaming.

BUT

> She could *hardly* keep from screaming.

NOT

> I was *not* doing *nothing* to her.

BUT

> I was doing *nothing* to her.

OR

> I was *not* doing anything to her.

NOTE:  **Double negatives** to emphasize the positive are allowable (but often are showy and artificial):

> It was *not for nothing* that she spent so many years in medical training.

■ **Exercise 21**

*Underline unacceptable forms of adjectives and adverbs, and write the correct forms. Write C by any correct sentence.*

1. The night was beautiful, and she smelled sweetly.

2. During that cold winter, birds sought frantic for food.

3. Computers sure are complicated.

4. Both actors performed awkward and tentative during the first act.

5. Then toward the end of the play, they acted their parts convincingly.

6. I was real pleased with their performance overall.

7. Think logical in times of confusion and pressure.

8. The acoustical situation was so bad that we could scarce hear the speaker.

9. We felt badly that we had to leave before the lecture was over.

10. Jennifer did volunteer work at the nursing home gladly and frequent.

■ **Exercise 22**

*Follow instructions for Exercise 21.*

1. Pluto is more further from the sun than is Earth.

2. When compared to Earth, Pluto is the smallest.

3. As far as scientists have determined, there is not nobody on Pluto.

4. Sir Malcolm Campbell's Blue Bird had a more unique design than any other racing car.

5. Katherine was the youngest of three daughters born into one of the most noblest families in England.

6. Of the four tours, the one to Brazil is the less expensive.

7. If you have to choose between being safe and being sorry, being safe is best.

8. Though several candidates are running, Leon seems the most qualified for the office.

9. Compared to, say, accounting, teaching offers the least pay.

10. You did not hardly even notice me.

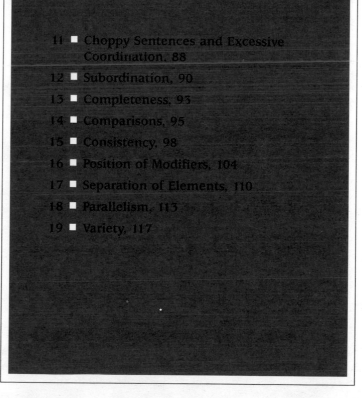

# Sentence
# Structure

# 11 Choppy Sentences and Excessive Coordination *chop/coord*

Do not present thought in brief, staccato sentences or in a string of independent clauses.

A group of short sentences or a number of independent clauses in the same sentence can obscure the relationship between ideas and create monotony.

## 11a Consolidate brief sentences in succession.

Brief sentences, one after the other, create **choppiness**. Ideas appear jumbled and fragmented; jerkiness and repetition destroy any chance of a graceful style.

CHOPPY

> The Valkyries appear in Norse poetry. They are large women. They have beautiful skin. They have golden hair. They wear helmets and breastplates. They carry spears and shields. They ride their horses through the sky.

Note the improvement when the sentences are consolidated.

IMPROVED

> The Valkyries of Norse poetry are large women with beautiful skin and golden hair. They wear helmets and breastplates and carry spears and shields as they ride their horses through the sky.

## 11b Do not string together short independent clauses.

In an attempt to avoid a choppy passage, some inexperienced writers insert a *but* or an *and* (or another coordinating conjunction) between the brief sentences and thus create **excessive coordination**.

EXCESSIVELY COORDINATED

> Mount Vesuvius is a volcano, and it is in Italy, and it is perhaps the most famous volcano in the world, for it has erupted several times, and it has caused great devastation, but it is also famous in a more positive way, for its slopes are extremely fertile, and some of the country's best wine comes from the grapes that grow there.

Skillful writers avoid the childish, repetitious style that results from excessive coordination. They vary their sentences by using phrases and dependent clauses that demonstrate the logical connections between ideas.

IMPROVED

> Italy's Mount Vesuvius is perhaps the most famous volcano in the world, for it has erupted several times with devastating results. In a more positive sense, it is famous for the wine made from grapes that grow on its fertile slopes

## ■ Exercise 1

*Rewrite the following sentences to avoid choppiness and excessive co-ordination.*

1. Rain fell most of the day. The rain suddenly stopped. The sun came out. The sun shone brightly. Elizabeth said this was remarkable. She said the rain had washed the sun.

2. The point-contact transistor was invented in 1948. It was invented by two scientists. Their names are John Bardeen and Walter H. Brattain. They worked at the Bell Telephone Laboratories.

3. The manta ray has a wide body, and it is flat, and it is also called a devilfish, and it is graceful.

4. Benjamin Franklin was an American, but he was at home wherever he went, and he gained wide popularity in France, but he was also well known in England.

5. Sharks are ferocious, and they attack many bathers each year, but they are just seeking food, and they seldom kill.

# 12 Subordination   *sub*

Use subordination to achieve proper emphasis and effective, varied construction.

Putting the less important idea of a sentence in a dependent clause emphasizes the more important thought in the independent clause. However, piling one dependent clause on top of another awkwardly stretches out sentences and obscures meaning.

## 12a   Express main ideas in independent clauses, less important ideas in dependent clauses.

An optimistic sociologist wishing to stress progress despite crime would write:

> Although the crime rate is high, society has progressed in some ways.

A pessimistic sociologist might wish the opposite emphasis:

> Although society has progressed in some ways, the crime rate is high.

**Upside-down subordination** results from placing the main idea of a sentence in a dependent clause. The writer of the following sentence wished to stress the insight and wisdom possessed by a young woman despite her youth.

> ┌─────────── *dependent clause* ──────────┐ ┌*independent*
> Although she possessed unusual insight and wisdom, she was still
>
> *clause* ┐
> young.

Unintentionally, the writer of the above sentence emphasized the person's youth. For the proper stress, the sentence should read:

> ┌──── *dependent clause* ────┐ ┌──── *independent clause* ────
> Although she was still young, she possessed unusual insight and
>
> ────┐
> wisdom.

## **12b**  Avoid excessive overlapping of dependent clauses.

Monotony and confusion can result from a series of clauses in which each depends on the previous one.

OVERLAPPING

> Pianos are instruments
>     ↳
>     that contain metal strings
>       ↳
>       that make sounds when struck by felt-covered hammers
>         ↳
>         that are operated by keys.

IMPROVED

Pianos are instruments containing metal strings that make sounds when struck by felt-covered hammers operated by keys.

## ■ Exercise 2

*Explain differences in meaning or emphasis in the following pairs of sentences.*

1. When life seems darkest, dawn appears.
   When dawn appears, life seems darkest.

2. Although she played the cello poorly, she was an effective legislator.
   Although she was an effective legislator, she played the cello poorly.

3. Even though the article was dull, the author made valuable comments.
   Even though the author made valuable comments, the article was dull.

4. While seeking fame, the writer found himself.
   While finding himself, the writer sought fame.

5. Although life is short, something can be accomplished.
   Although something can be accomplished, life is short.

## ■ Exercise 3

*Rewrite the following sentences to avoid excessive subordination.*

1. Numerous people who love animals are volunteering their time, which is precious, to their local humane societies that are trying to find homes for cats and dogs that are currently homeless.

2. A controversy that is raging centers on college football players who accept money and gifts, which is illegal, from agents, who are unscrupulous.

3. In a discovery that took place in 1939, construction workers found a cave that is near Monte Circeo, Italy, that had been sealed for fifty thousand years.

4. Inside the cave was a circle that was made of rocks that surrounded a skull, which had been broken, that was that of an ancient man.

5. Lobster Newburg is a dish that consists of cooked lobster meat, which is heated in a chafing dish that contains a special cream sauce.

# 13 Completeness   *compl*

Make your sentences complete in structure and thought.

Include all necessary words to make the parts of a sentence consistent and the meaning clear.

**13a** Do not omit a verb or a preposition that is necessary to the structure of the sentence.

OMITTED PREPOSITION

The apes were both attracted and suspicious **of** the stuffed animal placed in their midst. [The apes were attracted **of**?]

IMPROVED

The apes were both attracted **to** and suspicious **of** the stuffed animal placed in their midst.

BETTER

The apes were both attracted **to** the stuffed animal placed in their midst and suspicious **of** it.

OMITTED VERB

The cabinet *drawers* **were** open and the glass *door* shattered. [The door **were** shattered?]

VERB STATED

The cabinet *drawers* **were** open, and the glass *door* **was** shattered.

When the same verb form is called for in both elements, it need not be repeated:

To err is human; to forgive, divine.

**13b** Omission of *that* sometimes obscures meaning.

INCOMPLETE

The labor leader reported a strike was not likely.

COMPLETE

The labor leader reported **that** a strike was not likely.

# 14 Comparisons *comp*

Make comparisons logical and clear.

## 14a Do not make illogical comparisons.

Compare only similar terms. The following sentence illogically compares a baseball uniform and a player.

> Look at the difference between the way the pitcher is dressed and the batter.

A logical and clear comparison would link uniform with uniform or player with player.

> Look at the difference between the ways the pitcher and the batter are dressed.

OR

> Look at the difference between the pitcher and the batter.

## 14b Use the word *other* when it is needed in a comparison.

INCOMPLETE AND ILLOGICAL

> Alaska is larger in land area than any state.

COMPLETE

> Alaska is larger in land area than any *other* state.

# **14c**   Avoid awkward and incomplete comparisons.

AWKWARD AND INCOMPLETE

> The new cars appear to be as small if not smaller than last year's models. [*As small* requires *as,* not *than.*]

BETTER

> The new cars appear to be as small as last year's models if not smaller.

AWKWARD AND INCOMPLETE

> During much of the nineteenth century, Queen Victoria of Great Britain was one of the best known if not the best known woman in the world. [After *one of the best known,* the plural *women* is required.]

BETTER

> During much of the nineteenth century, Queen Victoria of Great Britain was one of the best-known women in the world if not the best known.

AMBIGUOUS

> After many years, Professor Drinkwater remembered me better than my roommate. [Better than he remembered my roommate, or better than my roommate remembered me?]

CLEAR

> After many years, Professor Drinkwater remembered me better than my roommate did.

OR

> After many years, Professor Drinkwater remembered me better than he did my roommate.

■ **Exercise 4**

*Correct any errors in completeness and comparisons.*

1. The adventurer found the buried treasure was but a myth.

2. Some wildflowers are as beautiful if not more beautiful than hothouse varieties.

3. The aorta is larger than any artery in the body.

4. Taxes are usually higher in urban areas than the country.

5. I have always and will always be true to you.

6. For good health, plain water is as good if not better than most other liquids.

7. The lighthouse stood as a symbol and guide to safety.

8. Cats eat more cheese than mice.

9. Statistics reveal that the average age of our nation's populace is and will continue to increase.

10. The king's daughters were ugly but the prince handsome.

■ Exercise 5

*Follow the instructions for Exercise 4.*

1. Some people believe the Bermuda triangle is more dangerous than any part of the world.

2. Copernicus discovered the sun is at the center of our solar system.

3. Birds seem to sing more beautifully in the early morning than the afternoon.

4. The editors say that the headlines have been written and the type set.

5. Shoppers standing in the checkout line are sometimes both curious and disapproving of the tabloids on the display racks.

# 15 Consistency *cons*

Write sentences that maintain consistency in form and meaning.

**15a** Avoid confusing or awkward shifts in grammatical forms.

## Shift in tense

PRESENT AND PAST

> The actors *rehearsed* and rehearsed, and then they finally *perform*.
> [Use *rehearsed . . . performed* or *rehearse . . . perform*.]

CONDITIONAL FORMS *(should, would, could)*

> Exhaustion after a vacation *could* be avoided if a family *can* plan
> better. [Use *could . . . would* or *can . . . can*.]

## Shift in person

> In department stores, *good salespersons* [**3rd person**] first ask whether
> a customer needs help or has questions. Then *they* [**3rd person**] do
> not hover around. If *you* [**2nd person**] do, *you* [**2nd person**] run the
> risk of making the customer uncomfortable or even angry [The third
> sentence should read, "If they do, they run. . . ."]

## Shift in number

> A *witness* may see an accident one way when it happens, and then
> *they* remember it an entirely different way when *they* testify. [Use
> *witness* and a singular pronoun or *witnesses* and *they . . . they*.]

## Shift in mood

*indicative*                   *imperative*
↓                       ↓

> First the job-seeker *mails* an application; then *go* for an interview. [Use
> *mails* and *he or she goes*.]

## Shift in voice

> The chef *cooks* [**active**] the shrimp casserole for thirty minutes, and
> then it *is allowed* [**passive**] to cool. [Use *cooks* and *allows*. See 5.]

### Shift in connectors

RELATIVE PRONOUNS

> She went to the chest of drawers *that* leaned perilously forward and *which* always resisted every attempt to open it. [Use *that . . . that.*]

CONJUNCTIONS

> The guests came *since* the food was good and *because* the music was soothing. [Use *since . . . since* or *because . . . because.*]

### Shift in use of direct and indirect discourse

MIXED

> The swimmer says that the sea is calm and why would anyone fear it?

CONSISTENT

> The swimmer says that the sea is calm and asks why anyone would fear it.

■ **Exercise 6**

*Correct shifts in the following sentences.*

1. Customers no longer seem to chat with each other in laundromats; you have to be more careful than in the past.

2. Success in the experiments will be achieved this time if the assistants would follow the instructions precisely.

3. Contact lenses that are well fitted and which are well cared for usually cause few problems.

4. Ulysses S. Grant was looking forward to the day of Lee's surrender, but when that day came, he experiences a migraine headache and suffers through the ceremony.

5. The smiling actor backed carefully down the steps, jumped on his horse, and rides off, leaving a cloud of dust behind.

6. Enthusiastic runners like to get out every day, but occasionally the weather in this region would be too bad even for the most dedicated.

7. A lawyer does not always enjoy a good reputation, for they are the subjects of many jokes.

8. Before one purchases an electrical appliance, you should find out how expensive it will be to operate.

9. It is wise to start a fire with a wood like pine, and then a heavier wood like oak is used.

10. Use pine first to start a fire since it ignites easily and because it will then make the oak burn.

**15b** Make subjects and predicates fit together logically. Avoid faulty predication.

Test a sentence by mentally placing the subject and predicate side by side. If they do not fit together (faulty predication), rewrite for clarity and logic.

FAULTY

> After eating a large meal is a bad time to go swimming. [Eating is not a time.]

LOGICAL

> The period just after eating a large meal is a bad time to go swimming. [*Period* and *time* match.]

FAULTY

> When crime statistics decrease causes some people to become careless. [*When* does not cause the carelessness.]

LOGICAL

> Decreasing crime statistics cause some people to become careless.

**15c** Avoid the constructions *is when, is where,* and *the reason . . . is because.*

Constructions with *is when, is where,* or *the reason . . . is because* are both illogical and ungrammatical.

FAULTY

> Tragedy *is where* a person of high standing falls to a low estate. [Tragedy is not *where*.]

LOGICAL

> Tragedy *occurs* when a person of high standing falls to a low estate.

FAULTY

> The reason why Macbeth falls is *because* he is overly ambitious.

LOGICAL

> The reason why Macbeth falls is *that* he is overly ambitious.

## ■ Exercise 7

*Correct any faulty constructions in the following sentences.*

1. Faulty predication is when the subject and verb do not fit together logically.

2. Because she is always late explains why she was fired.

3. The only time people should be late is if they cannot possibly avoid it.

4. The reason why so many houses in coastal Florida have flat roofs is because they endure the strong winds of hurricanes.

5. Volcanic action is why many islands in the South Seas have beaches with dark sand.

6. A pretzel is where a slender roll of dough sprinkled with salt is baked into the shape of a knot.

7. Some years ago, the growth in applications to graduate schools increased because of the recession.

8. An example of confusion is an intoxicated person in a house of mirrors.

9. Confusion is when you cannot think straight.

10. Apologetically, Edgar explained that the reason why he had not called was simply because he had been too busy.

# 16 Position of Modifiers  *dg/mod*

Attach modifiers clearly to the right word or element in the sentence.

A misplaced modifier can cause confusion or misunderstanding.

## 16a Avoid dangling modifiers.

A verbal phrase at the beginning of a sentence should modify the subject.

DANGLING PARTICIPLE

**Speeding** down the hill, several slalom *poles* were knocked down by the skier.

CLEAR

**Speeding** down the hill, the *skier* knocked down several slalom poles.

DANGLING GERUND

After **searching** around the attic, a *Halloween mask* was discovered. [The passive voice in the main clause causes the modifier to attach wrongly to the subject.]

CLEAR

After **searching** around the attic, *I* discovered a Halloween mask.

DANGLING INFINITIVE

**To enter** the house, the *lock* on the back door was picked. [*To enter the house* refers to no word in this sentence.]

CLEAR

To **enter** the house, *he* picked the lock on the back door.

DANGLING PREPOSITIONAL PHRASE

**After completing my household chores,** the *dog* was fed.

CLEAR

**After completing my household chores,** *I* fed the dog.

DANGLING ELLIPTICAL CLAUSE

**While still sleepy and tired,** the *counselor* lectured me on breaking rules.

CLEAR

**While I was still sleepy and tired,** the counselor lectured me on breaking rules.

NOTE:   Loosely attaching a verbal phrase to the end of a sentence is unemphatic.

UNEMPHATIC

> Tomatoes are a relatively recent addition to our tables, having been widely cultivated only in the last hundred years or so.

BETTER

> Tomatoes are a relatively recent addition to our tables; they have been widely cultivated only in the last hundred years or so.

Some verbal phrases that are **sentence modifiers** do not need to refer to a single word:

> *Strictly speaking,* does this sentence contain a dangling construction?
> *To tell the truth,* it does not.

## **16b** Avoid misplaced modifiers.

Placement of a modifier in a sentence affects meaning.

> He enlisted after he married *again.*
> He enlisted *again* after he married.

To prevent awkwardness or misreading, keep words, phrases, and clauses close to the words they modify.

MISLEADING

> Even on a *meager* sheriff's salary, a family may have some comforts.

CLEAR

> Even on a sheriff's *meager* salary, a family may have some comforts.

PUZZLING

> He admitted that he was the anonymous donor *in the letter.*

CLEAR

> He admitted *in the letter* that he was the anonymous donor.

**16c** Place limiting modifiers directly before the words they modify.

Qualifying or **limiting modifiers** are words like *only, just, even, almost, merely,* and *hardly.* Do not position such a modifier randomly; determine what word it goes with and place it immediately in front of that word. Careless positioning can cause awkwardness, create confusion, or change meaning. Because of the placement of *just* in the following sentences, two different meanings emerge.

> *Just* a word of encouragement is adequate.
> A word of encouragment is *just* adequate.

In the passage that follows, note misplaced limiting modifiers.

> Melvin *only* attended college for one reason: to meet girls. He
> *almost* failed all his subjects the first term; he *only* passed P.E. He
> *even* neglected to write home. Then he reformed and
> *merely* socialized on occasion until he improved his grades.

**16d** Avoid squinting modifiers.

A modifier placed between two words so that it can modify either word is said to squint.

SQUINTING
> Persons who exercise *frequently* will live longer.

CLEAR
> *Frequently* persons who exercise will live longer.

OR
> Persons who *frequently* exercise will live longer.

■ **Exercise 8**

*Correct faulty modifiers in the following sentences.*

1. A pessimist only thinks of the dark side of life.

2. Early one morning in Kenya, the photographer saw a white elephant while still in pajamas.

3. People who read often will develop a good vocabulary.

4. Looking through a magnifying glass, the flaw in the diamond appeared a dark spot.

5. Listeners often call in to radio talk shows who have strange beliefs.

6. While strolling near a railroad track, a satchel of money was discovered by three boys.

7. The review did not even include a brief summary of the novel's plot.

8. Those who lose sleep frequently cannot function properly.

9. Computers can solve problems in minutes with sophisticated programs that used to take months.

10. The carpenter inspected the board before sawing for nails.

■ **Exercise 9**

*Follow the instructions for Exercise 8.*

1. The restaurant offers meals for children that are inexpensive.

2. To diet effectively, calories must be counted.

3. Visitors can only enter some countries if they have the proper credentials.

4. Before landing at Plymouth, the Mayflower Compact was signed by the Pilgrims.

5. Serve some of the sandwiches now; keep the rest for the picnic in the refrigerator.

6. Deliberately and deviously the reporter said the senator had misled the press.

7. Without thinking, a daily routine can often be followed.

8. Desiring to calm the passengers, a witty remark was made by the captain.

9. Experiencing failure frequently leads to success.

10. Wishing to prevent the bends, the decompression chamber was entered by the diver.

# 17 Separation of Elements    *sep*

Do not divide any closely related parts of a sentence.

Placing words between subject and verb, between parts of a verb phrase, between verb and object, or between the two parts of an infinitive can result in awkwardness or confusion.

**17a** Avoid separation of subject and verb or verb and object.

AWKWARD

> Confucius, because of his many sayings dealing with the problems of life, has been recognized widely for his wisdom.

IMPROVED

> Because of his many sayings dealing with the problems of life, Confucius has been recognized widely for his wisdom.

AWKWARD

> The Alamo is known because it symbolizes the struggle to break Mexican rule as "the cradle of Texas liberty."

IMPROVED

> The Alamo is known as "the cradle of Texas liberty" because it symbolizes the struggle to break Mexican rule.

PUZZLING

> She is the man who owns the restaurant's wife.

CLEAR

> She is the wife of the man who owns the restaurant.

**17b** Do not divide a sentence with a quotation long enough to cause excessive separation.

AVOID

> Stephen Clarkston's view, "Politicians always create their own reality; during campaigns, they create their own unreality that is supposed to pass for reality," is cynical.

BETTER

> Stephen Clarkston's view is cynical: "Politicians always create their own reality; during campaigns, they create their own unreality that is supposed to pass for reality."

**17c** Generally, avoid split infinitives.

**Split infinitives** occur when a modifier divides *to* and the verb form, as in *to beautifully sing.*

AWKWARD

> ⌐ *splits the infinitive* ¬
> She was determined *to* **strongly and bravely** *face* the disease.

BETTER

> She was determined *to face* the disease strongly and bravely.

Some writers and readers object to split infinitives without exception. Others accept them when they do not create awkwardness. Because of possible opposition, simply avoid this form of separation.

■ **Exercise 10**

*Revise separations in the following sentences.*

1. Some early geographers failed to clearly and without bias see that the earth could not be flat.

2. Paul Christopher's finding on grave-robbings—"Most resurrectionists [body snatchers] in the United States avoided graveyards in highly populated areas in favor of pauper's plots"—is correct.

3. The Bible, because of its appeal to so many people over such a long period of time, has sold more copies than any another book in America.

4. Some infants have by the time they say their first words already cut several teeth.

5. Members at the first meeting of the aerobics class were asked to honestly and fairly estimate what their body weight should be.

# 18 Parallelism *paral*

Use parallel grammatical forms.

Elements in a sentence are *parallel* when one construction (or one part of speech) matches another: a phrase and a phrase, a clause and a clause, a verb and a verb, a noun and a noun, a verbal and a verbal, and so forth.

## 18a Use parallel constructions with coordinating conjunctions (*and, but, for,* etc.).

NOT PARALLEL

> *adjective*  *verb*
> ↓  ↓
> Sailing ships were *stately* and *made* little noise.

PARALLEL

> *adjectives*
> ↙  ↘
> Sailing ships were *stately* and *quiet*.

NOT PARALLEL

> *nouns*  *pronoun*
> ↙  ↘  ↓
> Young Lincoln read widely for *understanding, knowledge,* and *he* just liked books.

PARALLEL

> ┌──── *nouns* ────┐
> ↓
> Young Lincoln read widely for *understanding, knowledge,* and pleasure.

**18b** Repeat an article *(the, a, an)*, a preposition *(by, in, on,* etc.), the sign of the infinitive *(to)*, and other function words to clarify parallelism.

UNCLEAR

    The artist was *a* painter and sculptor of marble.

CLEAR

    The artist was *a* painter and *a* sculptor of marble.

UNCLEAR

    They passed the evening *by* eating and observing the crowds.

CLEAR

    They passed the evening *by* eating and *by* observing the crowds.

**18c** Use parallel constructions with correlatives *(not only . . . but also, either . . . or,* etc.)

NOT PARALLEL

                       *infinitive*         *preposition*
                          ↓               ↓
    Petroleum is used **not only** *to make* fuels **but also** *in* plastics.

NOT PARALLEL

        *verb*                          *preposition*
      ↓                            ↓
    **Not only** *is* petroleum used in fuels **but also** *in* plastics.

PARALLEL

                      *prepositions*
                    ↙            ↘
    Petroleum is used **not only** *in* fuels **but also** *in* plastics.

NOT PARALLEL

                    *adverb*          *pronoun*
                       ↓                 ↓
     The speeches were **either** *too* long, **or** *they* were not long enough.

NOT PARALLEL

          *article*                       *adverb*
             ↓                               ↓
     **Either** *the* speeches were too long **or** *too* short.

PARALLEL

                          *adverbs*
                        ↙         ↘
     The speeches were **either** *too* long **or** *too* short.

PARALLEL

     **Either** the speeches were too long, **or** they were too short.


## **18d** Use parallel constructions with *and who* and with *and which.*

Avoid *and who, and which,* or *and that* unless they are preceded by a matching *who, which,* or *that.*

NOT PARALLEL

     The position calls for a person with an open mind *and who* is cool-headed.

PARALLEL

     The position calls for a person *who* is open-minded *and who* is cool-headed.

PARALLEL

     The position calls for a person with an open mind and a cool head.

NOT PARALLEL

     A new dam was built to control floods *and that* would furnish recreation.

PARALLEL

A new dam was built to control floods and to furnish recreation.

PARALLEL

A new dam was built *that* would control floods *and that* would furnish recreation.

## ■ Exercise 11

*Revise sentences with faulty parallelism. Write C by any correct sentence.*

1. Lawrence of Arabia was a soldier, an adventurer, and he wrote a book.

2. From an early age the aspiring astronomer believed that his mission was to look farther into space than anyone else had looked and making new discoveries about distances in the universe.

3. Floppy disks are coated with magnetic material, covered by a protective jacket, and are for use in microcomputers.

4. At the time it appeared either impossible to get around the crowd or to go through it.

5. Children generally like bubble gum because it is appealingly packaged, sweet, and it lasts a long time.

6. The archaeologists decided to move to another site after spending three months digging, sifting, and unable to find anything of significance.

7. A young ballerina must practice long hours, give up pleasures, and one has to be able to take severe criticism.

8. The performance was long, boring, and made everyone in the audience restless.

9. Roaming through the great north woods, camping by a lake, and getting away from crowds are good ways to forget the cares of civilization.

10. The common cold brings on the following symptoms: a scratchy throat, a running nose, and you are generally fatigued.

# 19 Variety *var*

Do not write a passage made up of sentences unvaried in structure and word order.

Different kinds of sentences add a necessary variety to your writing. A passage composed of sentences all of the same kind is monotonous and dull, and it fails to indicate intricate relationships among ideas (see 11).

# **19a**  Do not overuse one kind of sentence structure.

Break the habit of following only one or two patterns in composing sentences. Vary your sentences among simple, compound, and complex patterns (see p. 26) and among loose, periodic, and balanced forms.

## *Loose sentences*

The most common kind of sentence, a **loose sentence,** makes its main point early and then adds further comment or detail.

LOOSE

> Conversation is necessary because it is a means by which people communicate ideas and ideals.

> *Uncle Tom's Cabin* is a very bad novel, having, in its self-righteous, virtuous sentimentality, much in common with *Little Women.*
>
> JAMES BALDWIN

## *Periodic sentences*

A **periodic sentence** withholds an element of the main thought until the end to create emphasis or suspense.

PERIODIC

> The most evident token and apparent sign of wisdom is a constant and unconstrained rejoicing.
>
> MICHEL DE MONTAIGNE

> Three people may keep a secret if two of them are dead.
>
> BENJAMIN FRANKLIN

## *Balanced sentences*

A **balanced sentence** has parallel parts that are similar in structure, length, and thought. Indeed, *balance* is simply a word for a kind of parallelism (see **18**).

BALANCED

Knowledge comes, but wisdom lingers.

ALFRED, LORD TENNYSON

That which is bitter to endure may be sweet to remember.

THOMAS FULLER

It is as easy to deceive oneself without perceiving it as it is difficult to deceive others without their perceiving it.

LA ROCHEFOUCAULD

A sentence can be balanced even if only parts of it are symmetrical. The balance in the next sentence consists of nouns in one group that are parallel to adjectives in the other.

Thus the Puritan was made up of two different men, the one all self-abasement, penitence, gratitude, passion; the other proud, calm, inflexible, sagacious.

THOMAS BABINGTON MACAULAY

Thus<br>
the Puritan<br>
was made up<br>
of two different men,

```
           the one --------- the other
all self-abasement, ------ proud,
       penitence, --------- calm,
       gratitude, -------- inflexible
       passion;-------- sagacious.
```

The following passage dealing with the use of the live oak tree in shipbuilding is monotonous because of its unrelenting use of loose sentences.

The wood from live oak trees was ideal for building sailing ships because of many characteristics. The shape of the tree's trunk and branches was one characteristic. The wood was also ideal for shipbuilding because it is dense, hard, strong, tough, and heavy. It is the heaviest of American woods. It is close-grained and therefore durable when in contact with soil and water. It has no characteristic odor. It did not cause nausea among people confined to the holds of ships. Other kinds of wood are subject to fungus attacks. The fungi rot the wood and cause a foul odor. This rotting odor became intolerable for anyone forced to stay below when the hatches were battened down.

Study the difference between the preceding paragraph and the one following, which exhibits variety in sentence structure.

In the building of sailing ships, several kinds of wood were acceptable, but that of the live oak tree was ideal. Many characteristics contributed to its suitability: the shape of the trunk and branches, its density, its hardness, its strength, and its heaviness. In fact, of all American woods, it is the heaviest. Since it is close-grained, it proved to be durable when in contact with soil and water. Because it has no characteristic odor and resists fungi, it did not cause nausea in people confined to the hold of ships with the hatches battened down. The timber of ships made of other woods rotted from the fungus attacks that produced foul odors, but the wood of the live oak tree remained free of fungi and odorless.

Adapted from LAURENCE C. WALKER
*Trees*

## 19b Vary the openings of sentences.

The most common order for a sentence is SUBJECT—VERB—OBJECT. This is a natural pattern in English, and it is fundamental to the language.

USUAL ORDER

     *subject*   *verb*           *object*   *modifiers* —————————————→
     ↓      ↓             ↓
She attributed these defects in her son's character to the general weaknesses of humankind.

If several consecutive sentences follow this order, however, they can become monotonous and take the edge off reader interest. Change the order occasionally. Study the following ways to create variety in word order.

SENTENCE BEGINNING WITH ADVERB
> *Quickly* the swordfish broke the surface of the water.

INVERTED SENTENCE BEGINNING WITH CLAUSE USED AS OBJECT
> *That the engineer tried to stop the train,* none would deny.

SENTENCE BEGINNING WITH PARTICIPIAL PHRASE
> *Flying low over the water,* the plane searched for the reef.

SENTENCE BEGINNING WITH DIRECT OBJECT
> These *defects* in her son's character she attributed to the general weaknesses of humanity.

SENTENCE BEGINNING WITH PREPOSITIONAL PHRASE
> *To the general weaknesses of humanity* she attributed the defects in her son's character.

## ■ Exercise 12

*Rewrite the following sentences and make them periodic. If you consider a sentence already periodic, write* P *by it.*

1. *Very* is a word much overused in the English language.

2. The story that Sir Isaac Newton formulated the law of gravity after watching an apple fall can be traced to one man, Voltaire.

3. A sense of humor is one quality that no great leader can be without.

4. Virginia Woolf was the daughter of Leslie Stephen, a well-known English writer.

5.  One machine, the computer, is revolutionizing science, stream-
    lining business practice, and influencing profoundly the
    writing habits of many authors.

■ **Exercise 13**

*Rewrite the following sentences to give them balanced constructions.
Place a B by any sentence that is already balanced.*

1.  It is wise to want to bring one's weight down to normal, but
    when the desire goes far beyond that, it is dangerous.

2.  The highest peak in North America is Mount McKinley; Mount
    Everest is the highest peak in the world.

3.  A trained ear hears many separate instruments in an orches-
    tra, but the melody is often all that is heard by the untutored.

4.  Realists know their limitations; romantics know only what
    they want.

5.  Our moods affect our perception of nature as our perception
    of nature affects our moods.

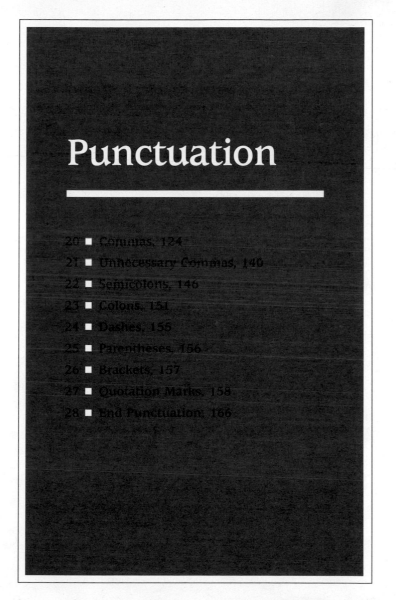

# Punctuation

# 20 Commas ,

Use commas to separate parts of sentences and to assure clarity.

**Commas** are internal marks of punctuation used to reflect the structure of a sentence and thereby make communication quick, clear, and graceful. Understanding a few principles will enable you to use commas effectively and confidently.

**20a** Use a comma to separate independent clauses joined by a coordinating conjunction (see p. 11).

> *coordinating conjunction*
> ↘
> Most children dislike coffee**,** yet many of them like it as they grow older.

> *coordinating conjunction*
> ↓
> Computers are useful and effective**,** but they will never replace books.

NOTE: The comma is sometimes omitted with a coordinating conjunction between the clauses when they are brief and there is no danger of misreading.

> The weather cleared and the aircraft departed.

**20b** Use commas between words, phrases, or clauses in a series.

> The closet contained worn clothes**,** old shoes**,** and dusty hats.

The final comma before *and* in a series is sometimes omitted.

> The closet contained worn clothes**,** old shoes and dusty hats.

But the comma must be used when *and* is omitted.

> The closet contained worn clothes, old shoes, dusty hats.

And it must be used to avoid misreading.

> An old chest in the corner was filled with nails, hammers, a hacksaw and blades, and a brace and bit.

Series of phrases or of dependent or independent clauses are also separated by commas.

PHRASES

> We hunted for the letter in the album, in the old trunks, and even under the rug

DEPENDENT CLAUSES

> Finally we concluded that the letter had been burned, that someone had taken it, or that it had never been written.

INDEPENDENT CLAUSES

> We left the attic, Father locked the door, and Mother suggested that we never unlock it again.

In a series of independent clauses, the comma is not omitted before the final element.

■ Exercise 1

*Insert commas where necessary.*

1. Necessity is the mother of invention or so it has been argued.

2. A good actor rehearses well enunciates clearly to be understood and practices effective timing.

3. Quilts were once made slowly by hand but now they are mass produced.

4. The sales manager and the trainee and a secretary visited branch offices in Pittsburgh in Dallas and in San Diego.

5. Rest is often necessary for the body needs time to recuperate.

6. The sensitive child knew that the earth was round but she thought that she was on the inside of it.

7. The lecturer exclaimed that Byron was exuberant that Wordsworth was inspired that Tennyson was gifted with music and that Sara Teasdale was profound.

8. For breakfast the menu offered only bacon and eggs toast and jelly and hot coffee.

9. Careless driving includes speeding stopping suddenly turning from the wrong lane going through red lights and failing to yield the right of way.

10. Do not underestimate your enemies for to do so could be a grave mistake.

**20c** Use a comma between coordinate adjectives not joined by *and*. Do not use a comma between cumulative adjectives.

**Coordinate adjectives** modify the noun independently.

COORDINATE

We entered a forest of tall, slender, straight pines.

Ferocious, alert, loyal dogs were essential to safety in the Middle Ages.

**Cumulative adjectives** modify the whole cluster of subsequent adjectives and the noun.

CUMULATIVE

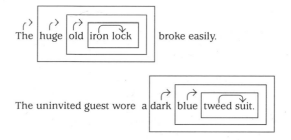

The huge old iron lock broke easily.

The uninvited guest wore a dark blue tweed suit.

Two tests are helpful.

### Test One

*And* is natural only between coordinate adjectives.

    tall *and* slender *and* straight pines
    ferocious *and* alert *and* loyal dogs

BUT NOT

    dark *and* blue *and* tweed suit

    huge *and* old *and* iron lock

### Test Two

Coordinate adjectives are easily reversible.

    straight, slender, tall pines

    loyal, alert, ferocious dogs

BUT NOT

    tweed blue dark suit

    iron old huge lock

The distinction is not always clear-cut, however, and the sense of the cluster must be the deciding factor.

She was wearing a full-skirted, low-cut velvet gown.

[a velvet gown that was full skirted and low cut, not a gown that was full skirted and low cut and velvet]

## ■ Exercise 2

*Insert commas where necessary. When in doubt, apply the tests described above. Write C by sentences that require no commas.*

1. Lhasa apso dogs have coarse straight hair.

2. From the ancient castle issued a shrill prolonged weird sound.

3. A prosecuting attorney needs a steady penetrating stern eye.

4. The red fox often has dusky reddish-brown fur.

5. The torn brown paper bag contained a fortune in cash.

6. Sarah Bernhardt was a beautiful talented creative actress.

7. That bitterly cold dark snowy day was etched in his memory.

8. The toga was a large loose outer garment worn by Roman citizens.

9. Rome was a dauntless highly organized widely feared power in ancient times.

10. A slight delicate agreeable scent arose from the garden.

## **20d** Use a comma after an introductory phrase or clause.

PHRASE

> With the most difficult part of the trek behind him, the traveler felt more confident.

CLAUSE

> When the most difficult part of the trek was behind him, the traveler felt more confident.

NOTE: A comma is always correct after introductory elements, but if the phrase or clause is short and cannot be misread, the comma may be omitted.

SHORT PHRASE

> After the ordeal the traveler felt more confident.

SHORT CLAUSE
>When day came the traveler felt more confident.

BUT

>*comma needed to prevent misreading*
>↓
>When sixty, people often consider retirement.

Introductory verbal phrases are usually set off by commas.

PARTICIPLE
>Living for centuries, redwoods often reach great heights.

INFINITIVE
>To verify a hypothesis, a scientist performs an experiment.

## **20e**  Use commas to set off nonessential elements.

A nonessential (nonrestrictive) element adds information that describes but does not alter the essential meaning of the sentence. When the modifier is omitted, the sentence loses some meaning but does not change radically.

NONESSENTIAL
>The painter's latest work, *a landscape*, has achieved wide acclaim.
>Salt, *which is plentiful in this country*, is still inexpensive.
>Languages contain abstract words, *which do not convey concrete images.*

NOTE:  *That* should never introduce a nonessential clause.

NOT
>Salt, *that* is plentiful in this country, is still inexpensive.

BUT
>Salt, *which* is plentiful in this country, is still inexpensive.

An essential (restrictive) element furnishes information that cannot be removed without radically changing the meaning of the sentence.

ESSENTIAL

> The Russian ruler *Nicholas* was married to Alexandra.
>
> Huge signs *that are displayed along a stretch of Highway 627* advertise a place called The Hornet's Nest.
>
> Words *that convey images* are important in poetry.

In all these sentences, the italicized expressions identify the words they modify; to remove the modifiers would change the meaning.

Some modifiers can be either essential or nonessential; use or omission of the commas changes the sense.

> The coin that gleamed in the sunlight was a Spanish doubloon. [There were several coins.]
>
> The coin, which gleamed in the sunlight, was a Spanish doubloon. [There was only one coin.]

## ■ Exercise 3

*Write* Y *by sentences that need commas in the places indicated,* N *by those that do not.*

1. Eric's only brother Jason decided to move to Australia.
   ∧          ∧

2. Barbers who are bald often recommend cures for baldness.
   ∧                ∧

3. The American composer George Gershwin wrote *Rhapsody in*
                          ∧                        ∧
   *Blue.*

4. Because of the extremely hot climate in the tropics ∧ many of

   the citizens wear large straw hats.

5. The most beautiful photograph ∧ a white bird against a darken-

   ing sky ∧ was made on a small island.

6. According to the latest national survey of gender in our coun-

   try ∧ women outnumber men.

7. All lawyers ∧ who are dishonest ∧ should be disbarred.

8. Only those ∧ who have the price of admission ∧ may enter.

9. Throughout a period of about twenty minutes ∧ the candidate

   never stopped talking.

10. During those long days of waiting and watching ∧ Elfin Grotte ∧ a

    teacher at the local community college ∧ never gave up hope

    that his wife would be found.

**20f** Use commas to set off sentence modifiers, conjunctive adverbs, and sentence elements out of normal word order.

Modifiers like *on the other hand, for example, in fact, in the first place, I believe, in his opinion, unfortunately,* and *certainly* are set off by commas.

Some poets, fortunately, do make a living by writing.

Thomas Hardy's poems, I believe, raise probing questions.

Commas are frequently used with **conjunctive adverbs** such as *accordingly, anyhow, besides, consequently, furthermore, hence, however, indeed, instead, likewise, meanwhile, moreover, nevertheless, otherwise, still, then, therefore, thus.*

BEFORE CLAUSE

The auditor checked the figures again; therefore, the mistake was discovered.

WITHIN CLAUSE

The auditor checked the figures again; the mistake, therefore, was discovered.

Commas **always** separate the conjunctive adverb *however* from the rest of the sentence.

The auditor found the error in the figures; however, the books still did not balance.

The auditor found the error in the figures; the books, however, still did not balance.

Commas are not used when *however* is an adverb meaning "no matter how."

However fast the hare ran, it could not catch the tortoise.

Use commas if necessary for clarity or emphasis when part of a sentence is out of normal order.

NOT NORMAL ORDER

Confident and informed, the young woman invested her own money.

OR

> The young woman, confident and informed, invested her own money.

NORMAL ORDER

> The confident and informed young woman invested her own money.

## 20g Use commas with degrees and titles and with elements in dates, places, and addresses.

DEGREES AND TITLES

> Sharon Weiss, M.A., applied for the position.
> Louis Ferranti, Jr., owns the tallest building in the city.
> Alphonse Jefferson, chief of police, made the arrest.

DATES

> Sunday, May 31, is her birthday.
> July 1994 was very wet. [Commas around 1994 are also acceptable.]
> July 20, 1969, was the date when a human being first stepped on the moon. [Use commas *before* and *after*.]
> He was born 31 December 1970. [Use no commas.]
> The year 1980 was a time of change. [Essential; use no commas.]

PLACES

> Cairo, Illinois, is my hometown. [Use commas *before* and *after*.]

ADDRESSES

> Write the editor of *The Atlantic*, 8 Arlington Street, Boston, Massachusetts 02116. [Use no comma before the zip code.]

## 20h Use commas for contrast or emphasis and with short interrogative elements.

> The pilot used an auxiliary landing field, not the city airport.
> The field was safe enough, wasn't it?

**20i** Use commas with mild interjections and with words like *yes* and *no*.

> Well, no one thought it was possible.
> No, it proved to be simple.

**20j** Use commas with words in direct address and after the salutation of a personal letter.

> Driver, stop the bus.
> Dear John,
> > It has been some time since I've written. . . .

**20k** Use commas with expressions like *he said, she remarked,* and *she replied* when used with quoted matter.

> "I am planning to enroll in Latin," she said, "at the beginning of next term."
> He replied, "It's all Greek to me."

**20L** Set off an absolute phrase with commas.

An **absolute phrase** consists of a noun followed by a modifier. It modifies the sentence as a whole, not any single element in it.

> ┌——*absolute phrase*——┐
> **Our day's journey over,** we made camp for the night.

> ┌——*absolute phrase*——┐
> **The portrait having dried,** the artist hung it on the wall.

**20m**   Use commas to prevent misreading or to mark an omission.

> After washing and grooming, the poodle looked like a different dog.
> When violently angry, elephants trumpet.
> Beyond, the open fields sloped gently to the sea.

*verb omitted*
↓

> To err *is* human; to forgive, divine. [Note that *is,* the verb omitted, is the same as the one stated.

■ **Exercise 4**

*Add necessary commas. If a sentence is correct as it stands, write* C *by it.*

1. To an ambitious person receiving much fame quickly is dangerous; receiving none at all devastating.

2. Geelong a city with an unusual name, is in South Victoria Australia.

3. A few hours before he was scheduled to leave the mercenary visited his father who pleaded with him to change his mind and then finally said quietly "Good luck."

4. Large winglike fins enable flying fish to glide through the air.

5. Inside the automobile looked almost new; however it was battered on the outside.

6. History one would think ought to teach people not to make the same mistakes again.

7. The Vandyke beard according to authorities was named after Sir Anthony Van Dyck a famous Flemish painter.

8. Measles is an acute, infectious viral disease that is recognizable by red circular spots.

9. Moving holidays from the middle of the week to Monday a recent national practice gives workers more consecutive days without work.

10. While burning cedar has a distinct strong odor.

■ **Exercise 5**

*Follow the instructions for Exercise 4.*

1. Seattle Pacific University according to its catalog was founded in Seattle Washington in 1891.

2. A small and affectionate dog of German breed the dachshund has a long body short legs and drooping ears.

3. A woman of spotless reputation and widely admired wisdom, Abigail Lindstrom R.N. will be greatly missed.

4. Most visitors who see the White House for the first time are surprised because it seems smaller than expected.

5. Discovered in 1799 the Rosetta stone a black basalt tablet offered the first clue fortunately to the deciphering of Egyptian hieroglyphics.

6. While the mystery writer was composing his last novel *The Tiger's Eye* he received a note warning him not to write about anyone he knew in the Orient.

7. The race being over the jockey who rode the winning horse turned to the owner and said "Now Mrs. Astor you have the money and the trophy."

8. The Levant is a region that includes Greece Turkey and Egypt.

9. Highway 280 known as "the world's most beautiful freeway" connects San Francisco and San Jose.

10. Of late streets of Hollywood California have been undergoing restoration.

# ■ Exercise 6

*Add necessary commas.*

1. A set of encyclopedias is helpful but a good dictionary is indispensable.

2. Yes I prefer small economical cars.

3. Attempting to save money as well as time some shoppers go to the grocery store only once a month; others however go almost daily.

4. The guest fell into a chair propped his feet on an ottoman placed his hands behind his head and yawned as the host glared at him.

5. However the travelers followed the worn outdated city map they always returned to the same place.

6. The last selection on the program a waltz by Strauss brought the most applause I believe.

7. Above the sky was peppered with tiny dark birds.

8. Ruth Friar Ph.D. was awarded her honorary degree on June 1 1947 in Fulton Missouri.

9. Yes friends the time has come for pausing not planning.

10. With a major snowstorm on the way people should stock up

on bread milk and eggs shouldn't they?

# 21 Unnecessary Commas *no ,*

Do not insert commas where they do not belong.

A page with a sprinkling of unneeded commas resembles a garment with unnecessary buttons scattered randomly throughout. When in doubt, do not use a comma.

## 21a Do not use a comma to separate subject and verb, or verb and object.

The mimosa tree**,** grows in several parts of the country. [subject and verb]
↑
*incorrect*

The geologist indicated that oil was present in the shale. [verb and object]
↑
*no comma*

## 21b Do not use a comma before a coordinating conjunction unless the conjunction joins independent clauses.

*no comma*
↓
The drama critic felt that the play would be engaging and that he would be able to write a favorable review. [independent clause followed by two dependent clauses]

*no comma*
↓
He left after the first act and tried to forget the play. [independent clause containing a compound verb]

**21c**  Do not use a comma to set off essential (restrictive) clauses, phrases, or appositives (see **20e**).

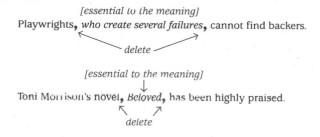

*[essential to the meaning]*
Playwrights, *who create several failures,* cannot find backers.
← *delete* →

*[essential to the meaning]*
↓
Toni Morrison's novel, *Beloved,* has been highly praised.
↖ ↗
*delete*

**21d**  Do not use a comma between adjectives that are not coordinate, between an adverb and an adjective, or between an adjective and a noun.

*no commas*
↓  ↓
A *dark green canvas* awning gave the cottage a look of comfort.
*[cumulative adjectives]*

*no comma*
↓
The badly damaged ship finally reached its destination.
 *[adverb]* *[adjective]*

*no comma*
↓
The stubborn, mischievous child would not respond to questions.
*[adjective]* *[noun]*

**21e** Do not use a comma after a coordinating conjunction.

*wrong*
↓

All entries must be submitted by Wednesday, and**,** the winner will be announced on Friday.

*wrong*
↓

Nor**,** will the judges allow late entries under any circumstances.

**21f** Do not use a comma before the subordinating conjunction *(after, although, because, before, if, since, unless, until, when, where)* when an adverbial clause follows an independent clause.

*no comma*
↓

We cannot leave today because the roads are impassable.

*no comma*
↓

Do not call unless an emergency arises.

**21g** Do not use a comma after the opening phrase of an inverted sentence (see p. 121).

*no comma*
↓

Close to the water's edge sits the house with seven gables.

**21h** Do not use a comma before the first item in a series or after the last.

*no commas*

The ancient chest contained rubies, emeralds, and pearls in abundance.

**21i**   Do not use a comma before *than* in a comparison.

*no comma*
↓
The pelican is a larger bird than the seagull.

**21j**   Do not use a comma after *like* or *such as*.

*no comma*
↓
The coin looked like a Spanish doubloon.

*no comma*
↓
A language such as Latin is no longer spoken.

**21k**   Do not use a comma with a period, a question mark, a dash, or an exclamation point.

*no comma*
↓
"Is there a motel nearby?" asked the traveler.

**21L**   Do not use a comma before parentheses.

*no comma*
↓
The celebration (for the mayor's re-election) lasted all day and all night.

■ **Exercise 7**

*Circle all unnecessary commas; be prepared to explain your decisions.*

1. A region in western Canada, the Klondike, once attracted, gold miners, adventurers, and opportunists, in abundance.

2. Tasmanian wolves, were once common, in Australia, but now they are confined to remote parts of Tasmania.

3. Restaurants, that serve excellent food at modest prices, are always popular among local people, though tourists seldom know about them.

4. In the wintry far north of Scandinavia, means of transportation like, snowshoes, skis, and sleds have to be used, because vehicles, with wheels, are impractical.

5. Ticket holders standing at the end of the line, worried that they would not find seats, or that the only available seats would be too far forward in the theater.

6. Although the composer had indicated that the piece was to be played adagio, (slowly), the conductor, sensing the audience's boredom, increased the tempo.

7. Yesterday, the sun shone, brightly.

8. Once, huge movie houses were fashionable, but now most of these palaces are like dinosaurs; that is, extinct, giants that are curious reminders, of the past.

9. A general practioner, is a physician, who does not specialize in any one field of medicine.

10. On the fifth day, came the word that a ship was in sight.

■ **Exercise 8**

*Circle all unnecessary commas. Write C by any correct sentence.*

1. The markup on handbags, was much more, than on shoes, on which the profit margin was slim.

2. Sales on sandals were seasonal, and unprofitable.

3. The most famous of all labyrinths, was that built by, Daedalus, for King Minos of Crete.

4. A filmy, cobweb wafted on the air, until a brisk wind carried it away.

5. The prairie dog is a small, quick rodent with a barking cry.

6. Marie Hautbois, who had just joined the orchestra, found an oboe in an antique shop and, she discovered that it was made in the seventeenth century.

7. She purchased the instrument, and told her conductor, Augustine Sey, of her great, fortune.

8. Upon first gazing at the Great Pyramid of Khufu, the tourist said that it was "prodigious," and that it was "inspiring beyond measure."

9. He found it grander, than the falls, that he had visited on the Zambezi River in Africa, or the Grand Canyon, (which he had seen the previous year), a popular tourist attraction in the United States.

10. Mammals such as, whales, and dolphins have flippers.

# 22 Semicolons ;

Use a semicolon between independent clauses not joined by coordinating conjunctions *(and, but, for, nor, or, so, yet)* and between coordinate elements with internal commas.

Omitting a semicolon between independent clauses may result in a comma splice or a fused sentence (see **2**).

**22a**  Use a semicolon between independent clauses not connected by a coordinating conjunction.

WITH NO CONNECTIVE

For fifteen years the painting was stored in the attic; even the artist forgot about it.

WITH A CONJUNCTIVE ADVERB

A specialist from the museum arrived and asked to examine it; *then* all the family became excited.

See **20f** for use of commas with conjunctive adverbs.

WITH A SENTENCE MODIFIER

The painting was valuable; *in fact,* the museum offered one hundred thousand dollars for it.

See **20f** for use of commas with sentence modifiers, such as *on the other hand, for example, in fact, in the first place.*

Notice the use of semicolons in the paragraph below, a lighthearted treatment of semicolons.

The semicolon tells you that there is still some question about the preceding full sentence; something needs to be added; it reminds you sometimes of the Greek usage. It is almost always a greater pleasure to come across a semicolon than a period. The period tells you that that is that; if you did not get all the meaning you wanted or expected, anyway you got all the writer intended to parcel out and now you have to move along. But with a semicolon there you get a pleasant little feeling of expectancy; there is more to come; read on; it will get clearer.

LEWIS THOMAS
*The Medusa and the Snail*

**22b**  Use a semicolon to separate independent clauses that are long and complex or that have internal punctuation.

In many compound sentences either a semicolon or a comma can be used.

COMMA OR SEMICOLON

*Moby-Dick*, by Melville, is an adventure story**,** [*or* **;**] and it is also one of the world's greatest philosophical novels.

SEMICOLON PREFERRED

Ishmael, the narrator, goes to sea, he says, "whenever it is a damp, drizzly November" in his soul**;** and Ahab, the captain of the ship, goes to sea because of his obsession to hunt and kill the great white whale, Moby Dick.

**22c**  Use semicolons in a series between items that have internal punctuation.

The small reference library included a few current periodicals, those most often read**;** a set of encyclopedias (the *Americana,* I believe)**;** several dictionaries, both abridged and unabridged**;** and various bibliographical tools.

**22d**  Do not use a semicolon between elements that are not grammatically equal.

*NOT* BETWEEN DEPENDENT AND INDEPENDENT CLAUSES

┌— *dependent clause*—┐ ┌————— *independent clause* —————┐
After it signaled the skiff**;** the coastguard cutter turned abruptly and left.
                          ↑
                *incorrect (use* **,***)*

┌── *dependent clause* ──┐ ┌──*independent clause*──┐
When the winds increase**;** the sea becomes rougher.
                         ↑
                *incorrect (use* **,** *)*

*NOT* BETWEEN A CLAUSE AND A PHRASE
┌──────────*clause*──────────┐ ┌──────── *phrase* ────────
Viewers of the eclipse were awestruck**;** having seen nothing like
┌────────┐
it before.                              ↑
                            *incorrect (use* **,** *)*

┌──────────── *clause* ────────────┐ ┌────── *phrase* ──────
The secrets of high productivity are simple**;** rising and retiring early
┌──────────────────────────────────┐
and wasting little time in between.        ↑
                            *incorrect (use* **,** *or* **:** *)*

■ **Exercise 9**

*Insert semicolons and change commas to semicolons where needed.*
*Write C by any correct sentence.*

1. Covered bridges once spanned many streams of New England

   they are rare now.

2. It was a warm, humid evening in, as I remember, mid-August

   and just outside the battered screen door, mosquitoes, thou-

   sands of them, gathered as crickets chattered madly.

3. The assignment was to seek out successful, experienced

   executives, to interview them, asking particular questions

about educational training, and to determine how important they themselves, looking back, considered their college training.

4. An advanced civilization is guided by enlightened self-interest; however, it is also marked by unselfish good will.

5. The Pawnees, who lived at one time in the valley of the Platte River in Nebraska, were once a numerous people; but their numbers diminished when the group, part of a confederacy of North American Plains Indians, moved from their ancestral homes to northern Oklahoma.

6. Irving was careful his friends said that he was indecisive.

7. The storm knocked down power lines, leaving the town dark uprooted trees, leaving the streets blocked and forced water over the levee, leaving the neighborhoods near the river flooded.

8. The gloomy hallway was narrow, long, and dark, and at the end of it hung a dim, obscure painting of a lean, haggard beggar in eighteenth-century London.

9. The soybean plant is now widely cultivated in America it is native, however, to China and Japan.

10. Fortunetelling still appeals to many people, they continue to patronize charlatans like palm readers.

# 23 Colons :

Use a colon as a formal mark of introduction.

Think of the colon as an arrow pointing to what follows it.

**23a** Use a colon after an independent clause that introduces a quotation.

AFTER AN INDEPENDENT CLAUSE

George Orwell predicted that personal writing and thus identity would be obliterated in 1984: [→] "The pen was an archaic instrument, seldom used even for signatures."

NOTE: After a brief independent clause, a comma rather than a colon may be used before the quotation.

Orwell writes, "The pen was an archaic instrument, seldom used even for signatures."

**23b** Use a colon after an independent clause that introduces a series of items.

BEFORE A SERIES

> An excellent physician exhibits four broad characteristics: [→]
> knowledge, skill, compassion, and integrity.

**23c** Use a colon after an independent clause that introduces an appositive.

An **appositive** is a word, phrase, or clause used as a noun and placed beside another word to explain, identify, or rename it.

> One factor is often missing from modern labor: pleasure in work.
> [Pleasure in work *is* the factor.]
>
> The author made a difficult decision: he would abandon the script.
> [To abandon the script *is* the decision.]

Frequently appositives are preceded by expressions like *namely* and *that is*.

> The new law calls for a big change: namely, truth in advertising. [Note
> that the colon comes **before** *namely,* not after.]

**23d** Use a colon between two independent clauses when one explains the other.

> Music communicates: it is an expression of deep feeling.

**23e** Use a colon after the salutation of a formal letter, between figures indicating hours and minutes, and in bibliographical entries.

Dear Dr. Tyndale**:**      *PMLA* 99 (1984)**:** 75
12**:** 15 P.M.              Boston**:** Houghton, 1929

**23f** Do not use a colon after a linking verb or after a preposition.

*NOT* AFTER LINKING VERB

*no colon*
↓

Some chief noisemakers **are** automobiles and airplanes.

*NOT* AFTER PREPOSITION

*no colon*
↓

His friend accused him **of** wiggling in his seat, talking during the lecture, and not remembering what was said.

■ **Exercise 10**

*Circle unnecessary colons, substituting other marks of punctuation when appropriate; add or substitute colons as necessary. Write C by sentences that are correct.*

1. According to historians, the first ascent of Mt. Everest was made by: Edmund Hillary and Tenzing Norkay.

2. Hillary and Norkay reached the top of Everest on May 29: 1953, at 6.50 in the morning.

3. Some climbers, however, dispute this historical account they believe the first ascent was accomplished by George Mallory.

4. In 1924: Mallory disappeared while attempting an ascent of that same Himalayan giant, Mt. Everest.

5. A few years ago another climbing party high on the mountain made an amazing discovery: a frozen body clad in climbing clothes of the 1920s.

6. Mallory's explanation of why he was trying to climb Everest has become famous, "Because it is there."

7. One of the greatest dangers in mountaineering is pulmonary edema, namely: the leakage of blood into the lungs.

8. Physicians on expeditions typically use two techniques to fight the symptoms; constant administration of oxygen and immediate evacuation to lower altitudes.

9. The route up the north face of Everest is: steep, unprotected, and subject to avalanches.

10. A successful climbing expedition depends on three crucial elements: physical skill, psychological strength, and an element of luck.

# 24 Dashes —

Use dashes to indicate sudden interruptions, to provide emphasis and readability, and to introduce summaries.

FOR SUDDEN INTERRUPTIONS

That September——no, it was late August——the rain fell unceasingly

FOR EMPHASIS OR READABILITY

Mako Hay——that unabashed hedonist——claims that the purpose of life is pleasure. [for emphasis]

Pleasing to the eye, though often flashy, paisley——an elaborate, intricate, colorful pattern——is still in demand. (Because of commas within the appositive, dashes make the sentence more readable.)

FOR SUMMARY

Hand fans, window fans, attic fans——all were ineffective that torrid summer.

NOTE: In typing, a dash is shown by two hyphens--like this-- usually with no space before or after.

# 25 Parentheses ( )

Use parentheses to enclose a loosely related comment or explanation, figures that number items in a series, and references in documentation.

FOR A COMMENT

> The frisky colt (it was not a thoroughbred) brought a good price at the auction.

A parenthetical sentence within another sentence has no period or capital, as in the example above. A freestanding parenthetical sentence requires parentheses, a capital, and a period.

*capital*        *period here*

↓          ↓

On that day all flights were on time. (The weather was clear.)

FOR FIGURES IN A SERIES

> The investor refused to buy the land because (1) it was too remote, (2) it was too expensive, and (3) the owner did not have a clear title.

FOR DATES OF BIRTH, DEATH, AND PUBLICATION

> Ludwig van Beethoven (1770–1827) was a German composer.
>
> *The Old Red House* (1860) was popular with young people in the nineteenth century.

FOR PAGE NUMBERS AND OTHER DOCUMENTATION

> *The Old Red House* begins, "I have been sitting in my vine-clad arbor" (3).
>
> Link does not agree (432).
>
> One critic disagrees (Link 432).

FOR CROSS-REFERENCES

> (See pp. 396–397 for further information on the use of parentheses in documentation.)

# 26 Brackets   [ ]

Use brackets to enclose interpolations within quotations.

> In the opinion of Arthur Miller, "There is no more reason for falling down in a faint before his **[**Aristotle's**]** *Poetics* than before Euclid's geometry."

Parenthetical elements within parentheses are indicated by brackets ([ ]). Try to avoid constructions that call for this intricate punctuation.

■ Exercise 11

*Supply dashes, parentheses, and brackets where needed.*

1.  Malcolm Cowley commented in *The View from 80* 1980 that "old people I wonder why have written comparatively little about the problems of aging."

2.  Malcolm Cowley 1898–1989 was surprised that old people did not write much about the difficulties of aging.

3.  Malcolm Cowley has written that "old people Cowley means the very old have written comparatively little about the problems of aging."

4. Malcolm Cowley points out 1 that many old people are still alert, 2 that they are highly sensitive to the difficulties of aging, and 3 that they do not write much about these problems.

5. Malcolm Cowley points out that many old people are still alert, that they are highly sensitive to the difficulties of aging, and that they possess an unusual degree of wisdom all factors that make them capable of writing about the phenomenon of growing old.

# 27 Quotation Marks    " "

Use quotation marks with the exact words of a speaker or writer and with some titles.

Double quotation marks (". . .") are preferred in American usage. Use single quotation marks for internal quotations (" '. . .' ").

## 27a  Use quotation marks to enclose direct quotations and dialogue.

DIRECT QUOTATION

The actress Lillie Langtry said, **"**The sentimentalist ages far more quickly than the person who loves his work and enjoys new challenges.**"**

CAUTION: Do *not* use quotation marks to enclose **indirect** quotations.

> The actress Lillie Langtry said that love of work and enjoyment of challenges keep one young far longer than sentimentalism.

DIALOGUE

> "What is fool's gold?" asked the traveler, who had never before tried prospecting.
>
> "Well," the old miner replied with a slight grin, "it's what makes fools of greenhorns."

In dialogue, each change of speaker begins a new paragraph.

**27b** Use quotation marks to enclose the titles of essays, articles, short stories, short poems, chapters (and other subdivisions of books or periodicals), dissertations (see pp. 390, 459), episodes of television programs, and short musical compositions.

> Kate Chopin's short work "The Story of an Hour" is about a woman's yearning for liberation.
>
> One chapter of *Walden* is entitled "The Beanfield." [For titles of books, see **30a**.]

**27c** Use single quotation marks to enclose a quotation within a quotation.

> According to one critic, "Dorothy Parker is merely pretending sentimentality when she swoons over 'One perfect rose.' "

**27d**  On your paper, do not use quotation marks around the title.

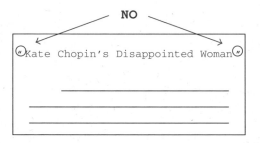

**27e**  Do not use quotation marks to emphasize or change the usual meanings of words or to justify slang, irony, or attempts at humor.

The beggar considered himself a "rich" man.

The old politician's enemies hoped that he would "croak."

Quotation marks do not give specialized or unusual definitions to words. They do not effectively add new meanings.

**27f**  Do not enclose a block (set-off) quotation in quotation marks.

Do not use quotation marks to enclose prose quotations that are longer than four lines. Instead, indicate the quotation by blocking—

indenting ten spaces from your left margin and single-spacing or double-spacing according to the preference of your instructor.

Unless your instructor specifies otherwise, poetry of four or more lines should be double-spaced and indented ten spaces. Retain the original division and format of the lines:

> A single flow'r he sent me, since we met.
> All tenderly his messenger he chose;
> Deep-hearted, pure, with scented dew still wet—
> One perfect rose.

NOTE:    Quotations of three lines of poetry or less may be written like the regular text—not set off. Use a **slash** ( **/** ) with a space before and after to separate lines.

> Dorothy Parker describes the flower as "Deep-hearted, pure, with scented dew still wet— **/** One perfect rose."

**27g** Follow established conventions in placing other marks of punctuation inside or outside closing quotation marks.

**Periods** and **commas** in American usage are placed *inside* closing quotation marks.

> All the students had read "Lycidas."
> "Amazing," the professor remarked.

NOTE:    With page numbers and other documentation, periods go outside quotation marks after the parentheses (see pp. 396, 460).

> The character echoed, with anticipation, "In the morning" (321).
> The author "intentionally delayed publication" (Lambert 32).

**Semicolons** and **colons** are placed *outside* closing quotation marks.

> The customer wrote that she was "not yet ready to buy the first edition"**;** it was too expensive.

A **question mark** or an **exclamation point** is placed *inside* closing quotation marks when the quotation itself is a direct question or an exclamation. Otherwise, these marks are placed *outside*.

> He asked, "Who is she**?"** [Only the quotation is a question.]
>
> "Who is she**?"** he asked. [Only the quotation is a question.]
>
> Did he ask, "Who is she**?"** [A quoted question within a question takes only one question mark—inside the quotation marks.]
>
> Did he say, "I know her"**?** [The entire sentence asks a question; the quotation makes a statement.]
>
> She screamed, "Run**!"** [Only the quotation is an exclamation.]
>
> Curse the man who whispers, "No"**!** [The entire statement is an exclamation; the quotation is not.]

After quotations, do not use a period or a comma together with an exclamation point or a question mark.

NOT
"When**?",** I asked.

BUT
"When**?"** I asked.

■ **Exercise 12**

*Insert double or single quotation marks where needed. Delete unnecessary quotation marks. Mark C by any correct sentence.*

1. A "tall" person can reach the higher shelves, whereas a "short"

   person cannot.

2. Eugene McCarthy made the following statement: An efficient bureaucracy is the greatest threat to liberty.

3. The greater the number of laws and enactments, in the words of the Chinese philosopher Lao-tzu, the more thieves and robbers there will be.

4. Robert Herrick's lyric poem Upon Julia's Clothes has been popular for centuries.

5.    Only in America, said the owner of the restaurant, could I have become wealthy in less than five years after immigrating.

   Yes, replied the customer, this country still offers opportunity for those who are willing to work hard.

6. An article in *Newsweek* entitled Re-Roofing the White House deals with the changes that always accompany new administrations.

7. To paraphrase what Robert Frost wrote, he took the road that had fewer travelers.

8. Do you believe Mark Twain was right when he said, Don't part with your illusions; when they are gone, you may still exist, but you have ceased to live?

9. The new homeowner said that "she could hardly believe how low the taxes were on her new house."

10. The newspaper article with the heading Bible Scholar Claims Discovery of Lost City reads as follows: Vendyle Jones said Thursday that the site lies at the center of a lost city. This city was, to quote Jones, greater than Troy, greater than Pompeii.

■ **Exercise 13**

*Add quotation marks where needed; circle unnecessary ones.*

1. "Failure is often necessary for humanity," the speaker said. Without failure, he continued, how can we retain our humility and know the full sweetness of success? For, as Emily Dickinson said, Success is counted sweetest / By those who ne'er succeed.

2. Madam, said the talent scout, I know that you think your daughter can sing, but, believe me, her voice makes the

strangest sounds I have ever heard. Mrs. Audubon took her daughter "Birdie" by the hand and haughtily left the room wondering "how she could ever have been so stupid as to expose her daughter to such a 'common' person."

3. Grandmother said, I'm going to teach you a short poem that I learned when I was a little girl. It's called The Wind, and it begins like this: I saw you toss the kites on high / And blow the birds about the sky.

4. The boy and his great-uncle were "real" friends, and the youngster would listen intently when the old man spoke. Son, he would say, I remember my father's words: You can't do better than to follow the advice of Ben Franklin, who said, One To-day is worth Two To-morrows.

5. To demonstrate that the English language is always "changing," our teacher said that we should all come up with a list of new expressions.

6. A recent report states the following: The marked increase in common stocks indicated a new sense of national security;

however, the report seems to imply "that this is only one of many gauges of the country's economic situation."

7. Chapters in modern novels rarely have any titles at all, especially "wordy" ones like the title of Chapter 51 in *Vanity Fair* (1848): In Which a Charade Is Acted Which May or May Not Puzzle the Reader.

8. On a hotel postcard sent to his "former" girlfriend, the young man marked an "X" on the picture of the building by the room where he was spending his "honeymoon" with his new bride.

9. This is where we are staying, he wrote. Wish you were here.

10. It was not at all "unusual" for the wanderer to stop busy passers-by on the street and ask them "why they were hurrying when life is so short."

# 28 End Punctuation   .?!

Use periods, question marks, or exclamation points to end sentences and to serve special functions.

## 28a Use a period after a sentence that makes a statement or expresses a command.

Some modern people claim to practice witchcraft.

Water the flowers.

The gardener asked whether the plant should be taken indoors. [This sentence is a statement even though it expresses an indirect question.]

## 28b Use periods after most abbreviations.

Periods follow such abbreviations as Mr., Dr., Pvt., Ave., B.C., A.M., Ph.D., e.g., and many others. In British usage, periods are often omitted after titles (Mr).

Abbreviations of government and international agencies often are written without periods (FCC, TVA, UNESCO, NATO, and so forth) Usage varies. Consult your dictionary.

A comma or another mark of punctuation may follow the period after an abbreviation, but at the end of a sentence only one period is used.

After she earned her M.A., she began study for her Ph.D.

But if the sentence is a question or an exclamation, the end punctuation mark follows the period after the abbreviation.

When does she expect to get her Ph.D.?

## 28c Use three spaced periods (ellipsis points) to show an omission in a quotation.

Note how the following quotation can be shortened with ellipsis points:

The drab, severe costumes of the Puritan settlers of New England, and their suspicion of color and ornaments as snares of the devil, have left their mark on the present-day clothes of New Englanders. At any large meeting, people from this part of the country will be dressed in darker hues—notably black, gray, and navy—often with touches of white that recall the starched collars and cuffs of Puritan costume. Fabrics will be plainer (though heavier and sometimes more expensive) and styles simpler, with less waste of material. Skirts and lapels and trimmings will be narrower.

ALISON LURIE
*The Language of Clothes*

ELLIPSIS POINTS TO INDICATE OMISSIONS

*ellipsis points not necessary here*
↓

Clothing "of the Puritan settlers of New England, and their suspicion of color and ornament **. . .**, have left their mark on the present-day clothes of New Englanders. At any large meeting, people from this part of the country will be dressed in darker hues **. . .** often with touches of white that recall the starched collars and cuffs of Puritan costume. Fabrics will be plainer **. . .** and styles simpler**. . . .**"
↑ ↑

*Use three ellipsis points for omissions within sentences.*
*Use period plus three ellipsis points for omission at end of sentence.*

NOTE: Although the comma after ellipsis points in the first sentence above could be omitted, it is permissible to include it since it completes a parenthetical element of the sentence.

**28d** A title does not end with a period even when it is a complete sentence, but some titles include a question mark or an exclamation point.

*Nobody Knows My Name*      *Who Is Angelina***?**
*Absalom, Absalom***!**      *Ah***!** *Wilderness*

**28e** Use a question mark after a direct question.

> Do teachers file attendance reports**?**
> Teachers do file attendance reports**?** [a question in declarative form]

Question marks may follow separate questions within an inter-rogative sentence.

> Do you recall the time of the accident**?** the license numbers of the cars involved**?** the names of the drivers**?**

NOTE:  A question mark within parentheses shows that a date or a figure is historically doubtful.

> Pythagoras, who died in 497 B.C. (**?**), was a philosopher.

**28f** Do not use a parenthetical question mark or excla-mation point to indicate humor or sarcasm.

NOT
> The comedy (**?**) was a miserable failure.

**28g** Use an exclamation point after a word, a phrase, or a sentence to indicate strong exclamatory feeling.

> Wait**!** I forgot my lunch**!**
> What a ridiculous idea**!**
> Stop the bus**!**

Use exclamation points sparingly. After mild exclamations, use commas or periods.

NOT

Well❗ I was somewhat discouraged❗

BUT

Well, I was somewhat discouraged.

### ■ Exercise 14

*Make two photocopies of a paragraph from an essay or a book. Submit both an unmarked copy and a copy in which you use ellipsis points to indicate omissions in the middle of a sentence and at the end of another sentence.*

### ■ Exercise 15

*Insert periods, question marks, and exclamation points where appropriate and delete any inappropriately used.*

In 1994 the annual Three Stooges Convention was held in Prevose, PA, during the month of July The distinguished (?) group consisted of no fewer than fifteen hundred admirers of the Stooges! Most of them are imitators as well as admirers, though one has to ask if they consider themselves serious impersonators? Hitting one another with foam-rubber sledgehammers and throwing cream pies, they did resemble the old Stooges, but their voice imitations (!) left much to be desired. Some simply exclaimed over and over, "Nyuk, nyuk, nyuk" Most of the impressions did not go much beyond such sounds. As they got to know one another, one was heard to ask another, "Are you the moron from Connecticut" The answer was an emphatic "Soitenly" Many of those present are known within the circle by one or the other names of the Stooges: Kenny "Mr Moe" Heath, George "Mr Larry" Likar, and Dave "Mr Curley" Knight. Actually "Mr Moe" Heath is the Rev Kenny Heath, who delivered the benediction at the banquet, a congenial and smashing (!) affair.

Adapted from JERRY ZEZIMA
Newspaper report, *The Stamford Advocate*

# Mechanics

---

# 29 Manuscript Forms, Business Letters, and Résumés *ms*

Follow appropriate manuscript form in papers, business letters, and résumés.

## 29a Follow standard practices in preparing the final copy of a paper.

The physical appearance of a paper demonstrates an author's pride in his or her work. Someone applying for a job would never show up in ragged jeans and a dirty T-shirt. Similarly, your written ideas should be presented in a neat and orderly format.

Always check with your instructors about their preferences for manuscript form. Most will require papers that have been typed or produced on a word processor.

### Computer printouts

Be sure the printer produces clean, dark copies with no smears. Avoid dot-matrix printers if possible. Be sure to tear off the hole-punched edges and separate the pages if the computer uses continuous-feed paper. Double-space throughout.

### Typing

Use a black typewriter ribbon on sturdy white paper, 8½ by 11 inches. Do not use onionskin, erasable, legal-sized, or colored paper. Double-space.

### Handwriting

Write legibly with dark blue or black ink. Use ruled white paper (8½ by 11 inches), skip every other line, and write on only one side of each sheet of paper. Do not use paper torn from a spiral-bound notebook.

## Margins

Use one-inch margins on each side and margins approximately an inch and a half deep at the top and bottom of the paper. Indent the first word of a paragraph five spaces. Indent set-off (block) quotations ten spaces.

## Spacing

Leave **one** space after end punctuation (period, question mark, exclamation point). Leave **one** space after commas, semicolons, colons, closing parentheses, and brackets. Leave no space on either side of a dash or a hyphen.

## Title and Name

If your paper does not have a separate title page, type your name, the name of the instructor, the course number, and the date on separate lines, all flush with the left margin. Double-space between the lines. Double-space again and center the title, and double-space between the title and the first line of the text. For examples of separate title pages, see pp. 401 and 463.

## Page Numbers

Use Arabic numerals (2, *not* II) in the upper right corner of the page. Numbering the first page is optional. Including your last name before the page number provides insurance in case pages are misplaced.

## *Example of correct manuscript form*

*double-space*  Mary Smith-O'Connor ←──────────── *identifying information flush left*

Professor Gates

History 101

15 December 1994

　　　　Immigration Quotas and the American Dream  ← *center the title*

*indent* ──→ Ever since the founding of the United States,
*5 spaces*
immigrants have landed on its shores hoping to find

a new way of life.  Those who arrived in New York

City after 1884 have been greeted by the towering

figure of the Statue of Liberty with its inviting

message: "Give me your tired, your poor, your hud-

←1"→ dled masses yearning to breathe free. . . ."  Some  ←1"→

came searching for religious freedom; others wanted

to escape political oppression; still others were

longing to escape the poverty of their homelands.

　　　　Immigration has flourished to such an extent,

however, that the United States has been faced with

difficult questions.  The notion of providing oppor-

tunity for the dispossessed is an integral part of

the country's political identity.  Yet should the

United States allow unlimited numbers of people to

↑
1"
↓

*name and page* ↓ ½" ↑
*flush right* → Smith-O'Connor   2

immigrate given today's increasingly limited land,

resources, and jobs?

*2 hyphens, no spaces for dash*↓
The first massive wave of immigration--occurring

in the nineteenth century--brought in primarily

Europeans: Irish, Scandinavians, Germans, Italians,

and Poles. By the 1920s, the United States had

enacted its first comprehensive immigration legisla-

tion and established specific geographical quotas.

Those decisions were significantly revised in 1965.

## 29b Follow accepted practices in preparing business letters.

In writing a business letter, follow conventional forms. Letters should be short, direct, and specific. Type or print out the letter; single-space it; double-space between paragraphs. Three standard formats are the **block form** (all elements flush with the left margin), the **modified block form** (return address, closing, and signature aligned at the right margin), and the **indented form** (modified block form with indented paragraphs). Samples of these forms appear on pp. 179, 180, and 181. Business letters are usually written on stationery measuring 8½ by 11 inches.

Try to learn the name *and* the title of the addressee. With names, use *Mr.* for men and *Ms.* for women (unless the recipient has indicated a preference for *Mrs.* or *Miss*).

Opinions vary about the best salutation when the writer does not know the name, title, or sex of the person being addressed. If the sex of the addressee is known but the name is not, use *Dear Sir* or *Dear Madam*. If the sex is not known, use *Dear Sir or Madam, Dear Human Resources Director* (using the title), or use no salutation at all.

## 29c Follow accepted practices in preparing a résumé.

A **résumé** is a brief, factual outline that presents a job applicant's qualifications for employment. It should preferably be no longer than one page. Résumés are usually arranged topically under headings (such as "Education," "Work Experience," or "Extracurricular Activities"), then chronologically within each of these categories.

All résumés should include your name, address, and phone number; employment objectives; education; past work experience; pertinent extracurricular activities; skills and interests (including languages and computer abilities); and a list of references or the address of the placement office of your college, which will provide your transcript and recommendations on request.

A sample résumé appears on p. 182.

## *Block form*

```
224 Leigh Avenue, Apt. B        ←————— address of writer
Seattle, WA 98119
May 23, 1995    ←——————————————————— date

Ms. Maria DeLillo   ←——————————————— name and title of addressee
Human Resources Director
Northwest Power Company
8825 South Fairfield Street
Bellingham, WA 98225    ←——————————— full address

Dear Ms. DeLillo:    ←——————————————— salutation and name (use a colon)
```

I am writing in response to the advertisement for a public relations
position that Northwest Power placed in today's <u>Seattle Times</u>.

In three weeks I shall graduate with a B.A. degree from the University
of Washington, with a major in English and a minor in Communications.
During my studies, I have taken courses in writing, editing, public
relations, and economics, and I have maintained a grade point average
of 3.75. In addition, last fall I worked as a student intern in public
relations for the City of Seattle. In this capacity, I wrote press
releases, edited newsletters, and helped to coordinate the neighborhood
festival program. I am enclosing a résumé giving a more detailed
summary of my background.

I believe that my writing and organizational skills, as well as my
training and experience, equip me well for the position you have open.
I will be at my present address through June 15, and you can reach
me by phone at 555-2121. If you wish, I shall be glad to come to
Bellingham for an interview.

```
Yours truly,    ←——————————————————— complimentary close

George M. Levine  ←——————— signature, handwritten

George M. Levine    ←——————————————— name, typed


Enclosure  ←————————————————————————— indication of material enclosed
```

## *Modified block form*

141 Oakhurst Drive, Apt. 2A
Singleton, OH 54567
March 1, 1995

Mr. Freeman O. Zachary, Manager
Personnel Department
Keeson National Bank
P. O. Box 2387
Chicago, IL 34802

Dear Mr. Zachary:

I am writing to ask whether you will have an opening this coming summer
for someone with my qualifications. I am finishing my sophomore year
at Singleton State College, where I intend to major in economics and
finance. I have been active in several extracurricular activities,
including the Spanish Club and the Singleton Players.

Although I have no previous experience in banking, I am eager to learn,
and I am willing to take on any duties you feel appropriate. Chicago
is my home town, so I am especially anxious to find summer employment
there. I will be in Chicago for spring vacation from March 20 to 26
and will be available for an interview. In addition, I shall be pleased
to furnish you with letters of recommendation from some of my professors
here at Singleton State and with a transcript of my college record. I
appreciate your consideration, and I look forward to hearing from you.

Sincerely yours,

*Audrey DeVeers*

Audrey DeVeers

---

Audrey DeVeers
141 Oakhurst Drive, Apt. 2A
Singleton, OH 54567

┌─────────┐
: Place :
: stamp :
: here :
└─────────┘

Mr. Freeman O. Zachary, Manager
Personnel Department
Keeson National Bank
P. O. Box 2387
Chicago, IL 34802

## *Indented form*

7315 Altamont Circle
Green Bay, WI 54311
July 27, 1995

Mr. Andrew Y. Liephart, President
Leakproof Plumbing and Heating Co.
837 Hasty Industrial Way
Green Bay, WI 54311

Dear Mr. Liephart:

On Friday, July 3, two of your employees installed a new electric
hot-water heater in my residence. In removing the old water heater, the
workers allowed water containing rust to spill out and stain the carpeting in my finished basement. They acknowledged the accident, but they
indicated that such occurrences are common and almost inevitable. I do
not accept this explanation, and I wish to point out that I signed no
statement relieving your company of liability incurred through damages
to my property.

I have had the carpeting professionally cleaned, and I feel that
your company should reimburse me for this expense. I enclose a copy of
the bill from Likenew Carpet Cleaners in the amount of $93.16. I expect
to receive a check from you in this amount so that this matter can be
settled without my having to take further action.

Sincerely yours,

*Lucille Navarro*

Lucille Navarro

Enclosure

## *Folding the letter*

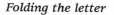

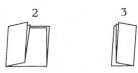

## *Résumé*

GEORGE M. LEVINE
2244 Leigh Avenue, Apt. B
Seattle, WA 98119
(206) 555-2121

OBJECTIVE
    To secure employment in the field of public relations

EDUCATION
    University of Washington, B.A. degree in English, June 1995.

WORK EXPERIENCE
    Student intern, the City of Seattle: worked as an assistant to
    Susan Miller, Director of Public Relations.  Duties included
    writing press releases, editing copy, answering phones, coordi-
    nating publicity and site arrangements for the neighborhood
    festival program. (June 1994-December 1994)

    Library assistant, Bellingham Public Library: checked out books,
    shelved books, processed new books.  (Summers 1991, 1992, 1993)

EXTRACURRICULAR ACTIVITIES
    Assistant Editor, <u>The Lake Union Review</u> (1993-94)
    Publicity coordinator, Young Life Club (1991-93)

SKILLS AND INTERESTS
    Computers: Word, Excel, Powerpoint, WordPerfect
    Language: Spanish
    Hobbies: Reading mysteries, kayaking, sailing

REFERENCES
    Available from the University of Washington Placement Office

# 30 Italics *ital*

Use italics for titles of separate publications (books, magazines, newspapers) and *occasionally* for emphasis.

Most books and magazines use mainly roman type: society. Type that slants to the right is called italic type: *society*. Many word-processing programs include italic type; in papers that are handwritten or typed, indicate italics by underlining: society.

**30a**  Italicize or underline titles of books (except the Bible and other sacred books and their divisions), periodicals, newspapers, long poems published separately, films, paintings, sculptures, musical compositions, television and radio programs, plays, and other long works published separately.

Be precise: underline initial articles *(A, An, The)* and any punctuation in a title.

BOOKS

Adventures of Huckleberry Finn (*not* The Adventures . . .)

An American Tragedy (*not* The American Tragedy)

*Not underlined:* the Bible, John 3:16, the Koran

PERIODICALS

The Atlantic and the American Quarterly

NEWSPAPERS

The New York Times or the New York Times or the New York Times

LONG POEMS

the Odyssey

FILMS

> *Rain Man*

MUSICAL COMPOSITIONS, RECORD ALBUMS, COMPACT DISCS

> *Beethoven's Mount of Olives* (*but not* Beethoven's Symphony No. 5)
>
> *The Best of Fleetwood Mac*

PLAYS AND MUSICALS

> *The Cherry Orchard*
>
> *Phantom of the Opera*

RADIO AND TELEVISION SERIES

> *All Things Considered*
>
> *Masterpiece Theatre*

**30b** Italicize or underline names of ships and trains.

> the *Queen Elizabeth II*    The U.S.S. *Hornet*    the *Zephyr*

**30c** Italicize or underline foreign words used in an English-language context, except for words that have become part of the English language.

Consult a dictionary to determine whether a word is still considered foreign or has become part of the English language.

> He claimed extravagantly to have been *au courant* since birth.
>
> The dandelion, of the genus *Taraxacum*, is a common weed.

BUT

Some words that may seem foreign have become a part of the English language.

> faux pas, amigo, karate

**30d** Italicize or underline words, letters, and figures being named.

> Remember to cross your t's. [or ts]
> Glee rhymes with flea.

NOTE: Occasionally quotation marks are used instead of underlining.

**30e** Avoid frequent italicizing or underlining for emphasis.

Numerous underlinings, dashes, and exclamation points on a page are distracting. Do not attempt to create excitement by overusing these marks.

**30f** Do not italicize or underline the title of your own paper.

■ **Exercise 1**

*Underline words as necessary. Place an X over words unnecessarily underlined. Place a C over words correctly underlined.*

1. The popular Broadway musical Cats is based on a book by T.S. Eliot called Old Possum's Book of Practical Cats.

2. The New Yorker reviewed the production last week along with the recent biographical study Eliot's New Life.

3. The reviewer insisted that Eliot be considered an American poet rather than a British one.

4. In many of Eliot's poems, he draws on material from the Bible.

5. His poem "Ash Wednesday" is included in the anthology The New Oxford Book of Christian Verse, edited by Donald Davie.

6. Director Terry Gilliam's film Brazil might be considered a cinematic rendition of Eliot's most famous long poem, The Waste Land.

7. The words burn and dry appear throughout the poem.

8. The professor read the short poem "Journey of the Magi" to the class while playing selections from Handel's Christmas oratorio, the Messiah.

9. Eliot may have traveled from the United States to England aboard the grand ocean liner the Queen Mary.

10. One painting in the National Gallery is titled Eliot and Pound in London.

# 31 Spelling *sp*

Spell correctly; make a habit of proofreading and checking words either with a spell-checker or in a dictionary.

Spelling is troublesome in English because many words are not spelled as they sound *(laughter, slaughter);* because some pairs and triplets sound the same *(capital, capitol; there, they're, their; to, too, two);* and because many words are pronounced with the vowel sound "uh," which gives no clue to spelling (sens*i*ble, cap*a*ble, defi*a*nt).

Many misspellings result from the omission of syllables in habitual mispronunciations *(accident-ly* for *acciden-tal-ly);* the addition of syllables *(disas-ter-ous* for *disas-trous);* or the changing of syllables *(prespiration* for *perspiration).*

## **31a** Spell-checking

Most word-processing programs include a spell-checker that will compare all the words in your paper against a dictionary that has been programmed into the computer. Such programs are good ways to do an initial check for spelling and proofreading errors, but they will not catch all of the errors.

In most programs, the computer will scroll through your manuscript and highlight words that do not match anything in the dictionary (names and unusual terms, even if properly spelled, will often be chosen). A spell-checker may also offer you several alternative spellings or words. You will have to make the final decision regarding each.

Spell-checkers do not determine whether a word is the right word for a particular sentence. You may, for example, accidentally type *the* for *they,* and the program will not catch the error. Similarly, if you confuse *its* and *it's, trail* and *trial,* or *to* and *two,* the program will not catch the error. Spell-checking must always be followed by proofreading.

## **31b** Proofreading

Many misspellings are simple typographical errors that can be caught by proofreading. Always read a manuscript through at least once just for proofreading.

Normally, when most people read, they take in several words at a time and jump from one word cluster to another. When proofreading, slow down your reading speed and try to focus on individual letters. Reading aloud can help you slow down. Other proofreading techniques include placing a ruler or index card under each line so that you focus on only one line. Read each paragraph or line word by word, forward or backward.

## **31c** Distinguishing homonyms

**Homonyms** are words that sound exactly like other words but have different spellings *(morning* and *mourning; there, their,* and *they're).* Different words that sound similar *(except* and *accept; loose* and *lose)* are near homonyms. Many spelling problems occur because of homonyms and near homonyms.

The Glossary of Usage (pp. 487–504) identifies and distinguishes between many common homonyms and near homonyms. Keeping a list of those words that frequently give you trouble and consulting the Glossary may help you avoid errors.

## **31d** Spelling strategies

English has no infallible guides to spelling, but the following suggestions are helpful.

## ie *or* ei?

Use *i* before *e*
Except after *c*
Or when sounded as *a*
As in n**ei**ghbor and w**ei**gh.

WORDS WITH *IE*

bel**ie**ve, ch**ie**f, f**ie**ld, gr**ie**f, p**ie**ce

WORDS WITH *EI* AFTER *C*

rec**ei**ve, rec**ei**pt, c**ei**ling, dec**ei**t, conc**ei**ve

WORDS WITH *EI* SOUNDED AS *A*

fr**ei**ght, v**ei**n, r**ei**gn

EXCEPTIONS TO MEMORIZE

**ei**ther, n**ei**ther, l**ei**sure, s**ei**ze, w**ei**rd, h**ei**ght

## *Drop final silent* e?

| DROP | | KEEP | |
|------|------|------|------|
| *When suffix begins with a vowel* | | *When suffix begins with a consonant* | |
| curse | cursing | live | lively |
| come | coming | nine | ninety |
| pursue | pursuing | hope | hopeful |
| arrange | arranging | love | loveless |
| dine | dining | arrange | arrangement |

TYPICAL EXCEPTIONS

coura**ge**ous
noti**ce**able
d**ye**ing (compare *dying*)
sin**ge**ing (compare *singing*)

TYPICAL EXCEPTIONS

awful
ninth
truly
argument

## *Change* y *to* i?

| CHANGE When y *is preceded by a consonant* | | DO NOT CHANGE When y *is preceded by a vowel* | |
|---|---|---|---|
| gully | gullies | valley | valle**y**s |
| try | tried | attorney | attorne**y**s |
| fly | flies | convey | conve**y**ed |
| apply | applied | pay | pa**y**s |
| party | parties | deploy | deplo**y**ing |

*When adding* -ing

| | |
|---|---|
| try | tr**y**ing |
| fly | fl**y**ing |
| apply | appl**y**ing |

## *Double final consonant?*

If the suffix begins with a consonant, do not double the final consonant of the base word *(man, manly)*.

If the suffix begins with a vowel:

| DOUBLE When final consonant is preceded by single vowel | | DO NOT DOUBLE When final consonant is preceded by two vowels | |
|---|---|---|---|
| | | despair | despairing |
| *Monosyllables* | | leer | leering |
| pen | pe**nn**ed | | |
| blot | blo**tt**ed | *Words ending with two or more consonants preceded by single vowel* | |
| hop | ho**pp**er | | |
| sit | si**tt**ing | | |
| | | jump | jumping |
| | | work | working |

| DOUBLE | | DO NOT DOUBLE | |
|---|---|---|---|
| | | *Polysyllables not accented on last syllable after addition of suffix* | |
| *Polysyllables accented on last syllable* | | | |
| defér | defe**rr**ing | defér | déference |
| begin | begi**nn**ing | prefér | préference |

| DOUBLE | | DO NOT DOUBLE | |
|--------|--------|--------|--------|
| omít | omi**tt**ing | devélop | devéloping |
| occúr | occu**rr**ing | lábor | lábored |

## Add s *or* es?

| ADD *S* | | ADD *ES* | |
|---------|--------|---------|--------|
| | | *When the plural is* | |
| *For plurals of most nouns* | | *pronounced as another syllable* | |
| girl | girl**s** | church | church**es** |
| book | book**s** | fox | fox**es** |
| | | *Usually for nouns ending in* | |
| *For nouns ending in* o | | o *preceded by a consonant* | |
| *preceded by a vowel* | | *(consult your dictionary)* | |
| radio | radio**s** | potato**es** | |
| cameo | cameo**s** | hero**es** | |
| | | BUT | |
| | | flamingos *or* flamingoes | |
| | | egos | |

NOTE:   The plurals of proper names are generally formed by adding *s* or *es (Darby,* the *Darbys; Jones,* the *Joneses).* Do not use an apostrophe to make a name plural.

| | PLURAL | NOT |
|-----|--------|-----|
| Ann | three Anns | three Ann's |

## Words frequently misspelled

Following is a list of over two hundred of the most commonly misspelled words in the English language.

| | | |
|-------------|--------------|-----------|
| absence | accumulate | advice |
| accessible | accuracy | advise |
| accidentally | acquaintance | all right |
| accommodate | acquitted | altar |

amateur
among
analysis
analyze
annual
apartment
apparatus
apparent
appearance
arctic
argument
arithmetic
ascend
athletic
attendance
balance
beginning
believe
benefited
boundaries
Britain
business
calendar
candidate
category
cemetery
changeable
changing
choose
chose
coming
commission
committee
comparative
compelled
conceivable
conferred
congratulations
conscience
conscientious
control
convenient
criticize
deceive

deferred
definite
democracy
description
desperate
dictionary
dining
disappearance
disappoint
disastrous
discipline
dissatisfied
dormitory
ecstatic
eighth
eligible
eliminate
embarrass
eminent
encouraging
environment
equipped
especially
exaggerate
excellence
exhilarate
existence
experience
explanation
familiar
fascinate
February
fiery
foreign
foreword
formerly
forty
fourth
frantically
fulfill or fulfil
generally
government
grammar
grandeur

grievous
guaranteed
gullible
harass
height
heroes
hindrance
hoping
humorous
hypocrisy
immediately
incidentally
incredible
independence
inevitable
intellectual
intelligence
interesting
irresistible
knowledge
knowledgeable
laboratory
laid
led
leisure
lightning
loneliness
maintenance
management
maneuver
manual
manufacture
marriage
mathematics
miniature
mischievous
mysterious
necessary
ninety
noticeable
occasionally
occurred
omitted
opportunity

optimistic
parallel
paralyze
pastime
perceive
performance
permanent
permissible
perseverance
personnel
perspiration
pervert
physical
picnicking
playwright
possibility
practically
precede
precedence
preference
preferred
prejudice
preparation
prescribe
prevalent
privilege
probably
professor

pronunciation
prophecy
prophesy
quantity
quiet
quite
quizzes
recede
receive
recognize
recommend
reference
referred
repetition
restaurant
rhythm
ridiculous
roommate
sacrifice
salary
schedule
secretary
seize
separate
sergeant
severely
shining
siege

similar
sincerely
sophomore
specifically
specimen
stationary
stationery
statue
studying
subtly
succeed
successful
summary
supersede
suppose
surprise
temperamental
tendency
their
thorough
through
vegetable
vengeance
villain
weird
writing
yield

## ■ Exercise 2

*Proofread the following passage; identify and correct all fifteen spelling or typing errors. Indicate with SC those errors that a spell-checker would not have identified.*

One of the especially accomplished woman of the twentieth century was the British author Dorothy L. Sayers. An excellant scholar of medeival literature and a respected translator of Dante,

Sayers today is perhaps best known as the author of a series of detective novels featuring the suave and distinguished Lord Peter Wimsey. After taking a degre at Oxford, Sayers worked as a copywriter in an advertising agency, inventing some of the most famous British ad campaignes of the 1920s. But eventually the success of her detective novels provided her with financial independance.

Asisted by his able manservent, Bunter, Lord Peter solves a vareity of intriguing mysteries—ranging from a mysterious death occuring at an advertiseing agency to the murder of his sister's finance. Lord Peter's most famous case, however, introduces him to Harriet Vane, a brilliant and passionete mystery writer who has been accused of poisoning the man with whom she lives. Peter solves the case, but falls in love with Harriet in a subplot that continues through several novels. Finaly, Harriet and Peter are engaged at the conclusion of *Gaudy Night;* during their honeymoon, recorded in *Busman's Honeymoon,* what should they discover but a corpse?

# 32 Hyphenation and Syllabication

Use a hyphen in certain compound words and in words divided at the end of a line.

Two related words not listed as an entry in a dictionary are usually written separately *(campaign promise)*.

**32a** Consult a dictionary to determine whether a compound is hyphenated or written as one or two words (see **37a**).

| HYPHENATED | ONE WORD | TWO WORDS |
|---|---|---|
| drop-off (noun) | droplight | drop leaf (noun) |
| white-hot | whitewash | white heat |
| water-cool | watermelon | water system |

**32b** Hyphenate a compound of two or more words used as a single modifier before a noun.

| HYPHEN | NO HYPHEN AFTER NOUN |
|---|---|
| She is a *well-known* executive. | The executive is *well known*. |
| Poe was a *nineteenth-century* writer. | Poe lived in the *nineteenth century*. |

A hyphen is not used when the first word of such a group is an adverb ending in *-ly*.

| HYPHEN | NO HYPHEN |
|---|---|
| a *half-finished* task | a *partly finished* task |

**32c** Hyphenate spelled-out compound numbers from *twenty-one* through *ninety-nine*.

**32d** Divide a word at the end of a line according to conventions.

### Monosyllables

Do not divide.

thought strength cheese

### Single letters

Do not put a one-letter syllable on a separate line.

NOT
a-bout might-y

### Prefixes and suffixes

May be divided.

separ-able pre-fix

Avoid carrying over a two-letter suffix.

NOT
bound-ed careful-ly

### *Compounds with hyphen*

Avoid dividing and adding another hyphen.

> self-satisfied

NOT
> self-satis-fied

■ **Exercise 3**

*Underline the correct form for the words indicated. Use a dictionary when needed.*

1. Many (summertime, summer-time, summer time) pleasures build (year-round, year round) memories.
2. The (thunderstorm, thunder-storm, thunder storm) drove many people off the (golfcourse, golf-course, golf course) into the (justopened, just-opened, just opened) (clubhouse, club-house, club house).
3. The (foxhound, fox-hound, fox hound) was (welltrained, well-trained, well trained) not to chase rabbits.
4. The (twentyone, twenty-one, twenty one) dancers did not know how to (foxtrot, fox-trot, fox trot).
5. Several words in the paper were hyphenated at the end of lines: a-round, almight-y, al-most, self-in-flicted, marb-le.

# 33 Apostrophes      '

Use the apostrophe for the possessive case of many nouns, for contractions, for omissions, and for some plurals.

**33a**  Use *'s* for the possessive of nouns not ending in *s*.

SINGULAR
  child**'s**, man**'s**, deer**'s**, lady**'s**

PLURAL
  children**'s**, men**'s**

**33b**  Use *'s* for the possessive of singular nouns ending in *s*.

  Charles**'s**, Watts**'s**, Dickens**'s**, waitress**'s**, actress**'s**

NOTE:  When a singular noun ending in *s* is followed by a word beginning with *s*, use only the apostrophe, not *'s*.

  the actress**'** success, Dickens**'** stories

**33c**  Use *'* without *s* to form the possessive of plural nouns ending in *s*.

  the Joneses**'** car, the Dickenses**'** home, waitresses**'** tips

**33d**  Use *'s* to form the possessive of indefinite pronouns.

  anybody**'s**, everyone**'s**, somebody else**'s**, neither**'s**

NOTE:  Use no apostrophe with personal pronouns like *his, hers, theirs, ours, its* (meaning "of it"). *It's* means "it is."

**33e**  Use *'s* with only the last noun for joint possession in a pair or a series.

the architect and the builder's plan [The two jointly have one plan.]
the architect's and the builder's plans [They have different plans.]

**33f**  Use ' to show omissions or to form contractions.

o'clock, jack-o'lantern
we'll, don't, can't, it's [meaning "it is"]

NOTE:  Be sure to distinguish between the following homonyms: *it's—its, you're—your, they're—their,* and *who's—whose.*

**33g**  Use *'s* to form the plural of acronyms and words being named.

three IRA's
six *and*'s

NOTE:  The plurals for numerals, letters, and years may be written with or without apostrophes as long as no confusion results (*u*'s not *u*s, *i*'s not *i*s). Be consistent.

three *7*'s or three *7*s, four *c*'s, or four *c*s, the 1960's or the 1960s.
the Roaring '20's or the Roaring '20s

■ **Exercise 4**

*Underline the words that contain correctly used apostrophes.*

1.  geese's honks, calves' hooves, ten dogs' yaps, a dogs' yap
2.  his', her's, it's (for *it is*), its'

3. Charles' SAT's, Les's ACTs, her father's IRA's
4. The childrens' service, the men's club, the man's hair.
5. Susan and Sam's patent (together they own one patent)
   Susan's and Sam's cars (each owns a car)
6. four oclock, four o'clock, four'o'clock
7. two *c*'s, two *ds*', three *is*
8. can't, couldnt, wouldn't
9. the Beatles's songs, the Beatles' album, the Beatles's fame
10. a childs' blocks, the two boys' toys, a girls' building block's

# 34 Capital Letters *cap*

Use a capital letter to begin a sentence and to designate a
proper noun (the name of a particular person, place, or
thing).

## 34a Capitalize the first word of every sentence.

The bat was made from the wood of an ash tree.

## 34b Capitalize first, last, and important words in titles, including the second part of hyphenated words.

> **A**cross the **R**iver and into the **T**rees
> "**T**he **M**an **A**gainst the **S**ky"
> "**A**fter **A**pple-**P**icking"
> "**W**hat **Y**ou **S**ee **I**s the **R**eal **Y**ou"

NOTE: Articles *(a, an, the),* short prepositions, and conjunctions are
not capitalized unless they begin or end a title.

**34c**   Capitalize the first word of a direct quotation introduced with a comma or a colon.

> Bruno Bettelheim states, "**T**he essential theme of 'Cinderella' is sibling rivalry."

NOTE:   Do not capitalize the first word of a quotation if you are including that word in the structure of your own sentence.

> Bruno Bettleheim claims that "the essential theme of 'Cinderella' is sibling rivalry."

**34d**   Capitalize titles preceding a name, following a name, or being used specifically as a substitute for a particular name.

> **P**rofessor Manoya
> Sandra Day O'Connor, **A**ssociate **J**ustice of the Supreme Court
> Dean Wilson approved the petition; the **D**ean is often lenient.

NOTE:   Do not capitalize a title that names an office if the title is not followed by a name (with the exception noted in **34e**).

> The professor gave the student the examination.
> A college president has more duties than privileges.
> A lieutenant usually leads a platoon.

**34e**   Capitalize the title of the head of a nation.

> the **P**resident of the United States
> the **P**rime **M**inister of Great Britain

## 34f    Capitalize proper nouns but not general terms.

| PROPER NOUNS | GENERAL TERMS |
|---|---|
| Plato, platonic, platonism | pasteurize |
| Venice, Venetian blind | a set of china |
| the West, a Westerner | west of the river |
| the Republican party | a republican government |
| the Senior Class of Ivy College | a member of the senior class |
| Clifton Street | my street, the street |
| the Mississippi River | the Mississippi and Ohio rivers |
| the Romantic Movement | the twentieth century |

## 34g    Capitalize words of family relationship that precede a name or that stand alone to represent a name.

After Father died, Mother and Aunt Alice carried on the business.

NOTE:  Do not capitalize words of family relationship that are preceded by a possessive pronoun or the word *the*.

After my father died, my mother and my aunt carried on the business.

## 34h    Capitalize the pronoun *I* and the interjection *O*.

My dear Romeo, O my dear Romeo, how I love you!

NOTE:  Do not capitalize the interjection *oh* unless it begins a sentence.

Oh, I like that film.

## 34i    Capitalize months, days of the week, and holidays.

April, Friday, the Fourth of July, Labor Day

NOTE: Do not capitalize seasons and numbered days of the month unless they name holidays.

spring, the third of July

## 34j Capitalize B.C. (used after numerals: 31 B.C.), A.D. (used before numerals: A.D. 33), words designating a deity, religious denominations, and sacred books.

in 273 **B.C.**

the **M**essiah, the **C**reator, **Y**ahweh, **A**llah, **C**atholic, **P**rotestant, **P**resbyterian, **M**uslim, **H**indu, **J**ewish, the **B**ible, the **K**oran

NOTE: Pronouns referring to God are usually capitalized.

Praise **G**od from **W**hom all blessings flow.

## 34k Capitalize names of specific courses.

Joseph registered for **B**iology 101 and **P**hilosophy 200.

NOTE: Do not capitalize subjects (other than languages) that do not name specific courses.

Joseph is taking French, biology, and philosophy.

## ■ Exercise 5

*Supply capitals as needed. Change capital letters to lowercase as necessary.*

1. Dr. Outback, an australian expert in Animal Behavior, lectures

   occasionally on Marsupial psychoses.

2. The Hostess, my aunt Zora, cried, "help yourself to the fried chicken," in a voice so shrill and strange that the dinner guests suddenly lost their appetites.

3. Captain Kaplan, united States army, arrived on wednesday to find that he was late for the tour of buddhist temples.

4. When she registered for Chemistry, amaryllis was told that she would need to take Algebra 101.

5. Susan Curall, m.d., attended the meeting of the American Medical association and returned home before thanksgiving day.

6. In the Twentieth Century, many of the qualities associated with people of the American south have disappeared.

7. Augustus Caesar, who was born in 63 b.c. and died in a.d. 14, is a character in Shakespeare's Tragedy *Antony and Cleopatra.*

8. Though the printer lived for a while on Magoni avenue, he moved to detroit last August.

9. The Salk Vaccine has all but eliminated Polio, according to an article in a Medical Journal.

10.  She wanted to become a Lawyer, she explained, because she

saw a direct connection between the Law and Morals.

# 35 Abbreviations and Symbols    *ab*

Avoid most abbreviations in formal writing.

## 35a  Spell out names of days, months, units of measurement, and (except in addresses) states and countries.

Friday (*not* Fri.)            pounds (*not* lbs.)

February (*not* Feb.)          Sauk Centre, Minnesota (*not* Minn.)

Do not use note-taking or shortcut signs such as *w/* for *with* or & for *and* in formal writing.

## 35b  Use only acceptable abbreviations.

BEFORE NAMES

Mr., Mrs., Ms., Messrs., Mmes., Dr., St. *or* Ste. (for *Saint,* not *Street*), Mt., Rev. (but only with a first name: *the Rev. Ernest Jones,* not *Rev. Jones*)

AFTER NAMES

M.D. (and other degrees), Jr., Sr., Esq.

FOR THE DISTRICT OF COLUMBIA

Washington, D.C.

FOR WELL-KNOWN ORGANIZATIONS (WITHOUT PERIODS)

    CIA, NAACP, FBI, IRS, NBC

NOTE:   Initially, identify less familiar organizations in parentheses; subsequently, use the acronym.

    The African National Congress (ANC) was banned as a political organization in South Africa for over fifty years. The ANC gained official recognition in 1992.

WITH DATES AND TIME

    B.C. and A.D. (with dates expressed in numerals, as *500 B.C.*), A.M. and P.M. or a.m. and p.m. (with hours expressed in numerals, as *4:00 A.M.* or *4:00 a.m.*)

## 35c   Use only acceptable symbols

PERCENT: %

    Write out the word *percent* in formal writing unless it appears in a table or graph. In technical, scientific, and business writing, the percent sign after a percentage is acceptable: *5%*.

DOLLAR SIGN: $

    Use a dollar sign only with specific dollar amounts: *$45.78* (but *several million dollars*).

CENT SIGN: ¢

    In formal writing, do not use a cent sign except in a table or a graph. Write *five cents*.

# 36   Numbers   *num*

Spell out numbers or use numerals where appropriate.

**36a**  Spell out numbers that can be written in one or two words. Use numerals for other numbers.

twenty-three, one thousand

123       $1^{13}/_{16}$       $1,001.00

NOTE:   In technical, scientific, and journalistic writing, writers often use numerals for all numbers above ten.

EXCEPTION:   Never use numerals at the beginning of a sentence. Spell out the number or recast the sentence.

**36b**  Be consistent with numbers in a sequence.

NOT

One polar bear weighed 426 pounds; another, 538 pounds; and the third, two hundred pounds.

BUT

One polar bear weighed 426 pounds; another, 538 pounds; and the third, 200 pounds.

**36c**  Use numerals for complete dates, street numbers, page and chapter references, percentages, and hours of the day used with A.M. or P.M.

USE NUMERALS

July 3, 1776 (*not* 3rd)
1010 State Street
See page 50.
He paid 15 percent interest.
The concert begins at 6 P.M.
   (or 6:00 P.M.)

SPELL OUT

the third of July
Fifth Avenue
The book has fifty pages.

The concert begins at
   six o'clock.

■ **Exercise 6**

*Place X by the following that are not acceptable in formal writing and correct them. Place C by those that are acceptable.*

1. On 25 December 1962, in Los Angeles, Calif.

2. Mr. Brown and Doctor Freidman

3. October twenty-third

4. four cents

5. Seventy five Jones Avenue

6. Devon Ave.

7. eight a.m.

8. Eng. 100 in the Dept. of English

9. 8 pounds, 5 oz.

10. two thousand and forty-five signatures

11. June 14th, 1994

12. $6.88 a pound

13. Peter & Paul

14. three little pigs

15. Thurs., February 14

16. page 24 in Chapter 12

17. Ken Griffey, Junior

18. 500 B.C.

19. the IRS

20. The jacket cost ten % less.

# Diction
# and Style

# 37   Diction   *d*

Use Standard English in formal writing.

Your **diction** (choice of words) should reflect a clear understanding of Standard English.

    **Standard English** is the generally accepted language of educated people in English-speaking countries. Though it varies in usage and in pronunciation from one country or region to another (indicated in dictionaries by labels such as *U.S.* or *Brit.*), it is substantially uniform. It is taught in schools and colleges and used in situations and writing where some formality is expected.

    **Nonstandard English** consists of usages, spellings, and pronunciations not usually found in the speech or writing of educated people. In a dictionary, labels indicate which words and expressions are nonstandard.

    **Informal** or **colloquial English** reflects the spoken language and is therefore more casual than is usually appropriate for college papers.

## 37a   Consult a good dictionary frequently.    *dic*

Dictionaries are excellent sources of information about language. They record past and current usage. Make a good dictionary your constant companion. Particularly useful at the college level are the following desk dictionaries:

*The American Heritage College Dictionary.* Boston: Houghton.

*The American Heritage Dictionary of the English Language.* Boston: Houghton.

*Merriam Webster's Collegiate Dictionary.* Springfield: Merriam.

*Random House Webster's College Dictionary.* New York: Random.

*Webster's New World Dictionary of American English.* New York: Prentice.

Become familiar with the various kinds of information offered in your particular dictionary. Read the front matter carefully and study its explanations. A typical entry from the *American Heritage College Dictionary* is represented on p. 215.

■ **Exercise 1**

*Read the preliminary pages in your dictionary, and study the usage labels (such as Mus., Math., Nonstandard, Informal, Slang, Vulgar, Regional, Chiefly Brit., and so forth) that it applies to particular words. Usage labels vary from one dictionary to another. Without using your dictionary, put by ten of the following words the labels, if any, that you believe they should have. Then look up each word, and determine which label it has. If a word is not labeled, it is considered formal.*

1. flat out (at top speed)
2. deck (data-processing cards)
3. gussied up
4. on deck (present)
5. movie
6. ain't
7. cram (to gorge with food)
8. tube (subway)
9. tube (television)
10. phony
11. bash (a party)
12. bash (a heavy blow)
13. enthused
14. pigeon (a dupe)
15. drip
16. freak (an enthusiast)
17. squeal (to betray)
18. deck (a tape deck)
19. wimp
20. on the level

**37b** Use words with precision; do not mistake one word for another. *den*

Make sure you know what a word means before you use it. Using the wrong word can create confusion or unintentional humor. The exact meanings of words are their **denotations,** their definitions given in dictionaries. Look up all words that you have any doubt about. Denotations are distinctive and precise.

MISUSED WORD

> Because of past problems, the young candidate is *venerable*.

WORD INTENDED

> Because of past problems, the young candidate is *vulnerable*.

*Venerable* means "old and respected," obviously not the meaning intended, which is "open to attack" *(vulnerable)*. Though the two words look and sound similar, they have almost opposite meanings.

MISUSED WORD

> A *tribe* of lions has a distinct social order.

WORD INTENDED

> A *pride* of lions has a distinct social order.

A *tribe* is a group of human beings. A group of lions is called a *pride*. Confusing the two could provoke a chuckle.

## Dictionary entry

Dictionaries provide the kinds of information indicated here in several ways. Many entries also include other information (synonyms, variants, capitalizations, and so on).

*Boldfaced main entry*
*correct spelling*

*Phonetic spelling for pronunciation*
*(dictionaries place a pronunciation key*
*at the bottom of each page)*

*Principal parts—for verbs*
*(past tense; present participle;*
*third person singular, present*
*tense)*

*Accented syllable*

*Meanings*
*of word*
*when used as*
*transitive verb*
*(tr.)*

*Dot indicating division*
*between syllables*

*Abbreviation for part of speech—verb*

↓   ↓  ↓ ↓   ↓   ↓

→ **con•nect** (kə-nĕkt′) *v.* **-nect•ed, -nect•ing, -nects.** — *tr.* **1.** To join or fasten together. **2.** To associate or consider as related. See Syns at **join. 3.** To join to or by means of a communications circuit. **4.** To plug in (an electrical cord or device) to an outlet. — *intr.* **1.** To become joined or united. **2.** To be scheduled so as to provide continuing service, as between airplanes. **3.** To establish a rapport or relationship; relate. **4.** *Sports.* To hit or play a ball successfully. [ME *connecten* < Lat. *cōnectere*: ← *cō-,* co- + *nectere,* to bind; see **ned-\*.**] — **con •nect′i •ble, con•nect′a•ble** *adj.* — **con •nec′tor, con•nect′er** *n.*

*Usage*
*label*

*Adjectives derived from*
*the entry word.*

*Nouns derived from*
*the entry word.*

*Meanings of word when used*
*as intransitive verb (intr.)*
*Note that the numbers start over*
*when the kind of verb changes.*

*Etymologies are given in square*
*brackets. Some dictionaries give*
*the etymology early in the entry.*

The italicized words in the following sentences are misused.

The new rocket was *literally* as fast as lightning. [*Figuratively* is intended.]

The captain of the team was *overtaken* by the heat. [The proper word is *overcome*.]

Words that are similar in sound but different in meaning can cause the writer embarrassment and the reader confusion.

A wrongful act usually results in a guilty *conscious*. [*conscience*]

The heroine of the novel was not embarrassed by her *congenial* infirmity. [*congenital*]

As the sun beams down, the swamp looks gray; it has only the unreal color of dead *vegetarian*. [*vegetation*]

A hurricane *reeks* havoc. [*wreaks*]

Displaced persons sometimes have no *lengths* with the past. [*links*]

■ **Exercise 2**

*Fifteen words are used incorrectly in the following passage. Underline them, and be prepared to discuss in class appropriate substitutes. Use your dictionary if necessary.*

Superstitions have always had a part in human behavior. The really elderly superstitions go far back, nearly to the gong of time when humanity was in its infamy. Human beings were like babies or toddies who cannot yet reason and who tremor with fear. Not understanding their fears, these early androids simply tried to find ways to feel more secure. They sensed that they could have control over their destinations if they could discover what was causing their malfeasance. They therefore began to denote certain actions and objects as the sources of their problems and to alternate their conduct accordingly. This revolved into a theory that they could change bad luck by an act of good luck. For instant, if they spilled salt, they could throw some over their shoulder and thus wart off unfavorable consequences. From this early way of thinking, superstition arose and flurried. Even today, people who are in many ways cognate deal with fears through superstition.

## 37c   Select words with desired connotations.   *con*

Many words carry **connotations:** associations or emotional over-tones that go beyond denotations. Effective writers carefully select words with connotations that will evoke particular emotional reactions. To suggest sophistication, the writer may mention a *lap dog;* to evoke the amusing or the rural, *hound dog.* To make a social or moral judgment, the writer may call someone a *cur.* Consider the associations aroused by *canine, pooch, mutt, mongrel, puppy,* and *watchdog.* Even some breeds arouse different responses: *bloodhound, shepherd, St. Bernard, poodle.*

    Words that are denotative synonyms may have very different connotative overtones. Consider the following:

> ambulance chaser—lawyer—attorney
> skinny—thin—slender

Be sure that your words give the suggestions you wish to convey. A single word with the wrong connotation can easily spoil a passage. President Lincoln once said in a speech:

> Human nature will not change. In any future great national trial, compared with the men of this, we shall have as weak and as strong, as silly and as wise, as bad and as good.

Notice how substituting only two words ruins the consistent high tone of the passage, even though the meanings of the substitutes are close to those of the original.

> **Folks** will not change. In any future great national trial, compared with the men of this, we shall have as weak and as strong, as silly and as wise, as **crooked** and as good.

■ **Exercise 3**

*Advertisers in particular are sensitive to the power of connotations. In class, explain the differences in the connotations of each word or phrase in parentheses in the advertising claims below.*

1. We have just received an unusually beautiful assortment of silk (lifelike, fake, artificial) flowers.
2. Our (boutique, store) has the largest selection in town of styles for (full-bodied, fat, stout) women.
3. (False teeth, Dentures) will sparkle after just one (scrubbing, brushing) with BriteUp.
4. For a (fun-filled, pleasurable) but (cheap, inexpensive, economical) tour of Europe, see Euro-Travel first.
5. If you (are a drunk, have a drinking problem), talk with a Curative Hospital (doctor, physician) today.
6. See Garden of Eden Florists for both potted plants and (severed, cut) flowers.
7. The Corner Shop carries (gifts, presents) for every occasion and specializes in (serving, waiting on) you promptly and courteously.
8. All (previously owned, used) automobiles at Olde Imports have been thoroughly (repaired, reconditioned).
9. The Empire Restaurant offers carefully (cooked, prepared) (food, meals, dishes) at reasonable prices.
10. We (guarantee, assure you) that our (toilets, restrooms, bathrooms) are the cleanest of any national chain of service stations.

## 37d Generally, avoid colloquialisms and contractions in formal writing.

A **colloquialism** is a word or expression suitable in everyday conversation and in some informal composition but inappropriate in college writing.

COLLOQUIAL
>*Fussing* at *kids* in public is a *no-no.*

APPROPRIATE IN FORMAL WRITING
>*Correcting children* in public is impolite.

A **contraction** is a word formed from two words by the omission of one or more letters. An apostrophe indicates the omission. Contractions are **informal**.

INFORMAL
>That *clerk's* the one who said *she'd* wait on me, but she *didn't.*

APPROPRIATE IN FORMAL WRITING
>That *clerk is* the one who said *she would* wait on me, but she *did not.*

■ **Exercise 4**

*In class, discuss usage in the following sentences.*

1. Teachers often get a bum rap even when they do a jam-up job.
2. The dispatcher reckoned that the semi would arrive soon, but he could not raise the driver on the radio.
3. If someone is putting you down, remain calm and simply tell him or her to chill out.
4. Why get bent out of shape over nothing? Play it cool.
5. That green stuff we had for dinner was really gross.
6. When the young stockbroker discovered the potential earning power of the mining company, he freaked out.
7. He couldn't do nothing without the supervisor eyeballing him and giving him a hard time.
8. The truckers heard on their ears that smokey was only three miles away on the blacktop.

9. The teller of the bank was instructed to tote the money back to the vault and to follow the boss to his office.
10. What cleared him was that the feds found out who really done the robbery.

## 37e  Generally, avoid slang.  *sl*

Slang expressions in student papers are usually out of place. They are particularly inappropriate in a context that is otherwise dignified.

> In the opinion of many students, the dean's commencement address *stunk*.
>
> When Macbeth recoiled at the thought of murder, Lady Macbeth urged him not to *chicken out*.

In addition, some slang words are so recent or localized that many people will have no idea of the meaning.

Although slang can be out of place, puzzling, or even offensive, it is at times effective. Some words that once were slang are now considered Standard English. "Jazz" in its original meaning and "dropout" were once slang; and because no other word was found to convey quite the same social meanings as "date," it is no longer slang. As a rule, however, avoid words considered slang. It is generally not as precise or as widely known as Standard English.

## 37f  Avoid dialect.  *dial*

Regional, occupational, or ethnic words and usages should be avoided in formal writing except when they are consciously used to give a special flavor to language.

There is no reason to erase all dialectal characteristics from talk and writing. They are a cultural heritage and a source of richness and variety.

**37g** Do not use a word as a particular part of speech when it does not properly serve that function.

Many nouns, for example, cannot also be used as verbs and vice versa. When in doubt, check the part of speech of a word in a dictionary.

NOUN FOR ADJECTIVE
*occupation* hazard

VERB DERIVED FROM NOUN FORM
*suspicioned*

VERB FOR NOUN
good *eats*

VERB FOR ADJECTIVE
*militate* leader

ADJECTIVE FOR ADVERB
sing *good*

**37h** Use correct idioms. *id*

An idiom is an accepted expression with its own distinct form and meaning. Language is rich in idiomatic expressions, some of which do not seem to make much sense literally but nevertheless have taken on specific meanings with time.

An onyx ring in the jewelry store *caught her eye*.
Though the gale was fierce, there was nothing else to do but *ride it out*.
The travel agent indicated that the ship would sail *Friday week*.

Many expressions demand specific prepositions; unidiomatic writing results when prepositions are misused. See also the list of two-word (phrasal) verbs on p. 511.

| UNIDIOMATIC | IDIOMATIC |
|---|---|
| according with | according to |
| capable to | capable of |
| conform in | conform to (or with) |
| die from | die of |
| ever now and then | every now and then |
| excepting for | except for |
| identical to | identical with |
| in accordance to | in accordance with |
| incapable to do | incapable of doing |
| in search for | in search of |
| intend on doing | intend to do |
| in the year of 1976 | in the year 1976 |
| lavish with gifts | lavish gifts on |
| off of | off |
| on a whole | on the whole |
| outlook of life | outlook on life |
| plan on | plan to |
| prior than | prior to |
| similar with | similar to |
| superior than | superior to |
| try and see | try to see |
| type of a | type of |

Dictionaries list many idiomatic combinations and give their exact meanings.

■ **Exercise 5**

*Select a verb that is the common denominator in several idioms, such as* carry *(carry out, carry over, carry through, and so forth). Look it up in a dictionary. Make a list, study the idioms, and be prepared to discuss them in class. Does the change of a preposition result in a different meaning? A slight change in meaning? No change?*

## ■ Exercise 6

*Underline the incorrect prepositions and supply the correct ones. Write*
C *by correct sentences.*

1. After studying with a tutor, the student who was behind in
   algebra was able to catch up to the rest of the class.

2. The jury acquitted him for the charge of loitering.

3. The aging baritone reacted favorably with the suggestion that
   he retire and surprised all members of the opera company.

4. The hiker said that he was incapable to going on.

5. The professor stated that his good students would conform
   with any requirement.

6. The employees complained with the manager because of their
   long hours.

7. Many vitamins are helpful to preventing diseases.

8. Fourteen new people were named to posts in the university
   administration.

9. Upon the whole, matters could be much worse.

10. Let us rejoice for the knowledge that we are free.

**37i** Avoid specialized vocabulary in writing for the general reader.    *tech*

All specialists, whether engineers, chefs, or philosophers, have their own vocabularies. Some technical words find their way into general use; most do not. The plant red clover is well known but not by its botanical name, *Trifolium pratense.*

Specialists should use the language of nonspecialists when they hope to communicate with general readers. The following passage, for instance, would not be comprehensible to a wide audience.

> The neonate's environment consists in primitively contrasted perceptual fields weak and strong: loud noises, bright lights, smooth surfaces, compared with silence, darkness and roughness. The behavior of the neonate has to be accounted for chiefly by inherited motor connections between receptors and effectors. There is at this stage, in addition to the autonomic nervous system, only the sensorimotor system to call on. And so the ability of the infant to discriminate is exceedingly low. But by receiving and sorting random data through the identification of recurrent regularity, he does begin to improve reception. Hence he can surrender the more easily to single motivations, ego-involvement in satisfactions.
>
> JAMES K. FEIBLEMAN<br>*The Stages of Human Life: A Biography of Entire Man*

Contrast the above passage with the following, which is on the same general subject but which is written so that the general reader—not just a specialized few—can understand it.

> Research clearly indicates that an infant's senses are functional at birth. He experiences the whack from the doctor. He is sensitive to pressure, to changes in temperature, and to pain, and he responds specifically to these stimuli. . . . How about sight? Research on infants 4–8 weeks of age shows that they can see about as well as adults. . . . The difference is that the infant cannot make sense out of what he

sees. Nevertheless, what he sees does register, and he begins to take in visual information at birth. . . . In summary, the neonate (an infant less than a month old) is sensitive not only to internal but also to external stimuli. Although he cannot respond adequately, he does take in and process information.

IRA J. GORDON
*Human Development: From Birth Through Adolescence*

The only technical term in the passage is *neonate;* unlike the writer of the first passage, who also uses the word, the second author defines it for the general reader.

Special vocabularies may obscure meaning. Moreover, they tempt the writer to use inflated words instead of plain ones—a style sometimes known as *gobbledygook* or *governmentese* because it flourishes in bureaucratic writing. Harry S. Truman made a famous statement about the presidency: "The buck stops here." This straightforward assertion might be written by some bureaucrats as follows: "It is incumbent upon the President of the United States of America to uphold the responsibility placed upon him by his constituents to exercise the final decision-making power."

## 37j Add new words to your vocabulary. *vocab*

Good writers know many words, and they can select the precise ones they need to express their meanings. A good vocabulary displays your intelligence, your education, and your talents as a writer.

In reading, pay careful attention to words you have not seen before. Look them up in a dictionary. Remember them. Recognize them the next time you see them. Learn to use them.

## ■ Exercise 7

*Underline the best definition.*

1. *colossal:* (a) colorful   (b) enormous   (c) terrible   (d) silent
2. *nominate:* (a) agree   (b) calculate   (c) think   (d) propose
3. *liaison:* (a) communication   (b) happy   (c) descendant   (d) loss
4. *armistice:* (a) seasoning   (b) musical composition   (c) hope   (d) truce
5. *toxin:* (a) shoe   (b) wagon   (c) poison   (d) champion
6. *charlatan:* (a) fraud   (b) chef   (c) North Carolinian   (d) specialist
7. *mediator:* (a) meat cutter   (b) reconciler   (c) mechanic   (d) test
8. *cryptic:* (a) mineral   (b) obscure   (c) tomb   (d) orderly
9. *lassitude:* (a) fatigue   (b) rope   (c) collie dog   (d) lecture
10. *gyration:* (a) fountain   (b) speech   (c) evasion   (d) revolving movement
11. *kindle:* (a) catch fire   (b) unearth   (c) taper   (d) hide
12. *harmonious:* (a) musical instrument   (b) in accord   (c) ugly   (d) decent
13. *perennial:* (a) relating to parents   (b) speedy   (c) pretty   (d) recurrent
14. *aspiration:* (a) medicine   (b) ambition   (c) drawing   (d) enlightenment
15. *theology:* (a) system   (b) theory   (c) study of religion   (d) political party
16. *quirk:* (a) fake   (b) store employee   (c) noise   (d) peculiarity
17. *fathom:* (a) sailor   (b) feather   (c) animal magnetism   (d) understand
18. *incorrigible:* (a) crated   (b) not reformable   (c) optimistic   (d) encouraging

19. *somber:* (a) lovely   (b) disputed   (c) gloomy   (d) Mexican hat
20. *referendum:* (a) vote   (b) reference work   (c) feast   (d) operation

■ **Exercise 8**

*Underline the best definition.*

1. *aromatic:* (a) easy   (b) automatic   (c) free   (d) fragrant
2. *heritage:* (a) inheritance   (b) germ   (c) upbringing   (d) nonconformity
3. *purge:* (a) article of clothing   (b) purify   (c) container   (d) happy sound
4. *tempest:* (a) pattern   (b) temporary worker   (c) storm   (d) large clock
5. *junket:* (a) trip   (b) useless item   (c) keg   (d) musician
6. *obviate:* (a) change   (b) alert   (c) manage   (d) make unnecessary
7. *dichotomy:* (a) elaborate game   (b) division   (c) diagram   (d) coloration
8. *plethora:* (a) fullness   (b) disease   (c) insect   (d) throat muscle
9. *recumbent:* (a) officeholder   (b) lying down   (c) appropriate   (d) false
10. *inimitable:* (a) matchless   (b) hateful   (c) fated   (d) final
11. *blunder:* (a) old rifle   (b) strong wind   (c) attempt   (d) mistake
12. *parapet:* (a) wall   (b) form of tax   (c) small bird   (d) medicine
13. *innocuous:* (a) injection   (b) harmless   (c) frightening   (d) preparation
14. *ordinance:* (a) heat   (b) opening   (c) regulation   (d) beginning

15. *whimsical:* (a) unpredictable   (b) wild   (c) clever   (d) dull
16. *salient:* (a) salty   (b) ill   (c) prominent   (d) rapid
17. *grovel:* (a) small stones   (b) smile   (c) creep   (d) tool
18. *debacle:* (a) laugh   (b) fiasco   (c) belch   (d) unruly discussion
19. *impediment:* (a) hindrance   (b) foot care   (c) undue haste   (d) favorite
20. *vivify:* (a) photograph   (b) quicken   (c) dramatize   (d) falsify

## ■ Exercise 9

*Place the number on the left by the appropriate letter on the right.*

| | | | |
|---|---|---|---|
| 1. ostracize | a. blockade |
| 2. transcend | b. glowing |
| 3. replica | c. tense |
| 4. vituperate | d. theory |
| 5. siege | e. free |
| 6. dwindle | f. ghost |
| 7. cow | g. companion |
| 8. preponderant | h. steal |
| 9. apparition | i. clear |
| 10. cohort | j. scold |
| 11. doldrums | k. banish |
| 12. vanguard | l. shrink |
| 13. taut | m. quickly passing |
| 14. disencumber | n. intimidate |
| 15. ignoble | o. despondency |
| 16. pellucid | p. exceed |
| 17. hypothesis | q. forefront |
| 18. pilfer | r. dominant |
| 19. transient | s. facsimile |
| 20. radiant | t. mean |

## ■ Exercise 10

*Underline the word or phrase that comes closest to defining the italicized word in each sentence.*

1. As time goes on, colleges appear to *proliferate* (grow richer, send more athletes into professional sports, grow in number, deteriorate).

2. *Holistic* (dealing with the whole, spiritual, scientific, relating to the mind) medicine is becoming widely accepted.

3. On that dark night, only the *silhouette* (front side, outline, long coat, profile) of the stranger could be seen.

4. The flamingo is a large *aquatic* (inedible, long-legged, domestic, water) bird.

5. Although no *philanthropist* (language specialist, generous giver, stamp collector, politician), the professor was highly respected.

6. The Santa Gertrudis is a breed of cattle highly valued for *hardiness* (muscularity, robustness, reproductive tendencies, strong will).

7. A *bilateral* (fair, economically feasible, carefully orchestrated, involving two sides) trade agreement could greatly aid both countries.

8. To be *pragmatic* (practical, prudish, pessimistic, stubborn) about it, experience is not always the best teacher.

9. *Expediency* (excessive costs, effectiveness, desire for change, emergency) often demands invention.

10. First acting slightly ill-natured, the waiter then responded with *alacrity* (insultingly, cheerful promptness, bitterness, lack of interest).

■ **Exercise 11**

*Underline the word or phrase that comes closest to defining the italicized word in each sentence.*

1. An *extemporaneous* (passionate, well-prepared, spontaneous, intelligent) speech is often effective.

2. Public speaking can be one of the most *horrendous* (rewarding, fascinating, dreadful, interesting) of experiences.

3. *Dogmatism* (stubborn opinionatedness, excessive love of animals, pettiness, shallowness) is the mark of a mind that oversimplifies.

4. One should speak with such volume that even those in the audience on the *periphery* (roof, with hearing aids, walking about, outer limits) can hear clearly.

5. *Candidness* (sweetness, expert photography, desire to run for office, openness) is nearly always appreciated.

6. On the other hand, *provincialism* (narrowness, immorality, arrogance, shyness) must be overcome.

7. A speaker who is obviously *disingenuous* (shabby, insincere, hostile, stupid) will seldom sway an audience.

8. Reaction to *taunts* (questions, comments, ridicule, laughs) reveals a speaker's degree of self-control.

9. The Tasmanian devil is a *carnivorous* (happy, aggressive, meat-eating, nocturnal) animal about the size of a badger.

10. From a *sequestered* (attractive, coastal, tiny, secluded) cottage, the voice of the famous tenor soared.

# 38 Style *st*

Express yourself with economy, clarity, and freshness.

**Style** refers to the way writers express their thoughts in language, the way they put words together. The modes in which similar ideas are expressed can have different effects, and many of the distinctions are a matter of style. Acquire the habit of noticing the personality of what you read and write, and work to develop your own style.

## **38a** Do not be wordy.    *w*

Conciseness increases the force of writing. A verbose style is flabby and ineffectual. Do not pad your paper merely to obtain a desired length.

USE ONE WORD FOR SEVERAL

> The love letter was written by somebody who did not sign a name. [13 words]

> The love letter was anonymous [*or* not signed]. [5 or 6 words]

USE THE ACTIVE VOICE FOR CONCISENESS (See 5)

> The truck was overloaded by the workmen. [7 words]

> The workmen overloaded the truck. [5 words]

DELETE UNNEEDED WORDS

> Another element that adds to the effectiveness of a speech is its emotional appeal. [14 words]

> Emotional appeal also makes a speech more effective. [8 words]

AVOID CONSTRUCTIONS WITH *IT IS . . .* AND *THERE ARE . . .*

> *It is* truth *that* will prevail. [6 words]
> Truth will prevail. [3 words]

> *There are* some conditions *that* are satisfactory. [7 words]
> Some conditions are satisfactory. [4 words]

DO NOT USE TWO WORDS WITH THE SAME MEANING (tautology)

> basic and fundamental principles [4 words]
> basic principles [2 words]

AVOID BUREAUCRATIC PHRASES

> at that point in time [5 words]
> at that point [3 words]

> early on [2 words]
> early [1 word]

Study your sentences carefully, and make them concise by using all the preceding methods. Do not, however, sacrifice concreteness and vividness for conciseness and brevity.

CONCRETE AND VIVID

> At each end of the sunken garden, worn granite steps, flanked by large magnolia trees, lead to formal paths.

EXCESSIVELY CONCISE

> The garden has steps at both ends.

■ **Exercise 12**

*Express the following sentences succinctly. Do not omit important ideas.*

1. The kudzu plant is a plant that was introduced into America from Japan in order to prevent erosion and the washing away

of the land and that has become a nuisance and a pest in some areas because it chokes out trees and other vegetation.

2. The cry of a peacock is audible to the ear for miles.

3. The custom that once was so popular of speaking to fellow students while passing by them on the campus has almost disappeared from college manners and habits.

4. There are several reasons why officers of the law ought to be trained in the law of the land, and two of these are as follows. The first of these reasons is that police officers can enforce the law better if they are familiar with it. And second, they will be less likely to violate the rights of private citizens if they know exactly and accurately what these rights are.

5. Although the Kentucky rifle played an important and significant part in getting food for the frontiersmen who settled the American West, its function as a means of protection was in no degree any less significant in their lives.

6. Some television programs, especially public television programs, assume a high level of public intelligence and present their shows to the public in an intelligent way.

7. It is possible that some of the large American chestnut trees survived the terrible blight of the trees in the year of 1925.

8. The plane was flown by a cautious pilot who was careful.

9. It is a pleasure for some to indulge in eating large quantities of food at meals, but medical doctors of medicine tell us that such pleasures can only bring with them unpleasant results in the long run of things.

10. The essay consists of facts that describe vividly many aspects of the work of a typical stockbroker. In this description the author uses a vocabulary that is easy to understand. This vocabulary is on neither too high a level nor too low a level, but on one that can be understood by any high school graduate.

# **38b** Avoid repeating ideas, words, or sounds. *rep*

Unintentional repetition is a mark of bad style. Avoid by omitting words and by using synonyms and pronouns.

REPETITIOUS

> The history of human flight is full of histories of failures on the part of those who have tried flight and were failures.

IMPROVED

> The history of human flight recounts many failures.

Do not revise by substituting synonyms for repeated words too often.

WORDY SYNONYMS

>   The history of human flight is full of stories of failures on the part of those who have tried to glide through the air and met with no success.

Do not needlessly repeat sounds.

REPEATED SOUNDS

>   The biologist again *checked* the *charts* to determine the *effect* of the poison on the *insect*.

IMPROVED

>   The biologist again studied the charts to determine the effect of the poison on the moth.

NOTE:   Effective repetition of a word or a phrase may unify, clarify, or create emphasis, especially in aphorisms or poetry.

>   *Searching* without knowledge is like *searching* in the dark.
>   "Beauty is truth, truth beauty." [John Keats]

■ **Exercise 13**

*Rewrite the following passage. Avoid wordiness and undesirable repetition.*

One of the pleasing things that all peoples of all lands have always enjoyed since the earliest dawning of civilizations is the pleasure of listening to the melodious strands of music. Music, wrote some great poet, can soothe the savage beast and make him or her calm. A question may arise in the minds of many, however, as to whether some of our members of the younger generation may not be exposing themselves too frequently to music they like and listen to almost constantly.

To me, the answer to the above question is just possibly in the affirmative, and the evidence that I would give would be in the form of a figure that we see nearly every day. This is the figure, usually a young person but not always, of a person with a radio or stereo headset who is listening to music while doing an activity that used to occupy all of one's attention and time. In the library of a college such a figure is now at times to be seen looking through books with a headset on his or her head. Some of these music lovers seem incapable of even taking a walk to enjoy the beauty of nature without their headsets.

If music is good and if we enjoy it, the reader may now ask, what is wrong with listening to music more frequently than we used to be able to listen to music thanks to the present modern improvements and developments in the electronic medium? It may seem old-fashioned and out-of-date to object to headsets just because we did not once have them and they are relatively new on the scene. What is important, here, however, is the point that those who spend so much of their time with headsets on are missing something. They may be getting a lot of music, but they are missing something. What they are missing can be classified under the general heading of other important and pleasurable sensory experiences that they could be experiencing, such as hearing the sounds that birds make and just enjoying the quiet that morning sometimes brings. In addition to this, it is hard to think while music is being piped into the ear of the listener. We lose time, therefore, that should by all rights be reserved for thinking and contemplation, which is to say, that time when we all develop and mature intellectually. Music, all kinds of music, even rock music, has an important place in the lives of humankind, but let us not so fall in love with its seductive appeal that beckons us that we let it intrude upon the territory of other valuable and essential experiences.

## 38c   Avoid flowery language.   *fl*

Flowery language is wordy, overwrought, and artificial. Often falsely elegant, it calls attention to itself. In the hope that such language will sound deep and wise, some inexperienced writers substitute it for naturalness and simplicity.

| PLAIN LANGUAGE | FLOWERY LANGUAGE |
| --- | --- |
| the year 1991 | the year of 1991 |
| now | at this point in time |
| lawn | verdant sward |
| shovel | simple instrument for delving into Mother Earth |
| a teacher | a dedicated toiler in the arduous labors of pedagogy |
| reading a textbook | following the lamp of knowledge in a textual tome |
| eating | partaking of the dietary sustenance of life |
| going overseas | traversing the ever-palpitating deep |

## 38d   Be clear.   *cl*

A hard-to-read style annoys and alienates. Do not take for granted that others will be patient or understanding enough to pore over your writing to determine your meaning. If you have any doubt that a sentence or a passage conveys the precise meaning you intend, revise and clarify. Nothing is more important in writing than making sense. Below are a few suggestions about communicating more clearly.

### Write specifically.

Abstract writing may communicate little information or even cause misunderstanding. Specific words say what you think. Their meaning is not ambiguous.

NOT SPECIFIC

> There was always a certain something in her personality that created a kind of positive effect.

SPECIFIC

> Her optimism always conveyed a sense of hope and joy.

### Give concrete examples.

Concrete examples often pierce to the heart of meaning, whereas vague and abstract writing ineffectually ranges around the periphery.

CONCRETE

> Her bright, quick smile and her musical way of saying "Good morning!" conveyed a sense of hope and joy.

### ■ Exercise 14

*The following ten words or phrases are general. For any five of them substitute four specific and concrete words. Do not let your substitutions be synonymous with each other.*

EXAMPLE

> tools      claw hammer, screwdriver, monkey wrench, saw

1. associations
2. rural (or urban) buildings
3. people who are failures
4. works of art
5. terrains
6. nuts (or vegetables or meats or breads)
7. missives
8. educational institutions above the second year of college
9. foot races
10. ways to walk

■ **Exercise 15**

*Write your personal definition of one of the following abstract terms in a paragraph of about two hundred words. Give concrete instances from your experience.*

1. education
2. family
3. maturity
4. pride
5. prejudice

**38e** Avoid triteness and clichés. Strive for fresh and original expressions.  *trite*

Clichés are phrases and figures of speech that were once fresh and original but have been used so much that they have lost their effectiveness. A stale phrase leaves the impression of a stale mind and a lazy imagination. Develop a sensitivity to clichés, and reject them for fresher writing.

Study the following twenty phrases as examples of triteness. Avoid pat expressions like these:

| | |
|---|---|
| words cannot express | method in their madness |
| each and every | straight from the shoulder |
| Mother Nature | first and foremost |
| sober as a judge | hard as a rock |
| other side of the coin | in the final analysis |
| slowly but surely | felt like an eternity |
| in this day and age | the bottom line |
| few and far between | all walks of life |
| last but not least | easier said than done |
| interesting to note | better late than never |

■ **Exercise 16**

*Underline the clichés in the following sentences.*

1. Once in a blue moon, Professor Alhambra tells a joke that is funny, but then he spoils it by laughing like a hyena.
2. International Bugging Machines is on the cutting edge of technology; therefore, its stock is expected to soar through the roof.
3. It goes without saying that the value of a college education cannot be measured in money, but tuition is as high as a kite.
4. As he looked back, the farmer thought of those mornings as cold as ice when the ground was as hard as a rock and when he shook like a leaf as he rose at the crack of dawn.
5. The survivor was weak as a kitten after eight days on the ocean, but in a few days he was fit as a fiddle.
6. The brothers were as different as night and day, but each drank like a fish.
7. The heiress claimed that she wanted to marry a strong, silent type, but she tied the knot with her hairdresser.
8. It was raining cats and dogs, but still she managed to look as pretty as a picture.
9. Surely it is possible to keep abreast of the times without giving up the tried and true values of the past.
10. If you pinch pennies now, you can enjoy the golden years with more peace of mind.

**38f**  Use fresh and effective figures of speech. Avoid mixed and inappropriate figures.    *fig*

Figures of speech compare one thing (usually abstract) with another (usually literal or concrete). Mixed figures associate things that are not logically consistent.

MIXED

> These corporations lashed out with legal loopholes. [You cannot *strike* with a *hole*.]

Inappropriate figures of speech compare one thing with another in a way that violates the mood or the intention.

INAPPROPRIATE

> Shakespeare is the most famous brave in our tribe of English writers. [It is inappropriate and puzzling to compare Shakespeare to an Indian brave and a group of writers to an Indian tribe.]

Use figurative comparisons (of things not literally similar) for vivid explanation and for originality. A simile, a metaphor, or a personification gives you a chance to compare or to explain what you are saying in a different way from the sometimes prosaic method of pure statement, argument, or logic.

METAPHORS (implied comparisons)

> Though calm without, the young senator was a volcano within.

> Old courthouse records are rotting leaves of the past.

SIMILES (comparisons stated with *like* or *as*)

> Though calm without, the young senator was like a volcano within.

> Old courthouse records are like rotting leaves of the past.

## ■ Exercise 17

*Explain the flaws in these figures of speech.*

1. The new race car flew once around the track and then limped into the pit like a sick horse into its stable.

2. The speaker's flamboyant oration began with all the beauty of the song of a canary.

3. Crickets chirped with the steadiness of the tread of a marching army.

4. The warm greetings of the students sounded like a horde of apes rushing through a jungle.

5. He nipped the plan in the bud by pouring cold water on all suggestions.

■ **Exercise 18**

*Compose and bring to class two fresh and appropriate figures of speech.*

■ **Exercise 19**

*Find two figures of speech in your reading, and show how they are mixed or especially appropriate or inappropriate.*

# Accurate Thinking and Writing

# 39  Accuracy and Logic   *log*

Accuracy and logic are indispensable for a sound and convincing argument. Do not talk or write in a way that misleads or gives false information or conclusions. Present facts without error, and reason logically. Admit uncertainties, and take into account valid opposing views or evidence.

**Logical argument** is either **inductive** or **deductive**.

## Inductive method

An inductive argument cites information derived from experiments, examples, cases, related facts, observations, statistics, and so forth to prove a certain conclusion or to establish a principle. This is the scientific method of reasoning, used in many kinds of writing. Inductive argument presents a hypothesis and then cites concrete and specific evidence to prove the reliability of the hypothesis. Trial lawyers often use the inductive method to present cases. From the quality of the evidence the jury determines which argument is likely to reflect the truth just as the reader measures the effectiveness of evidence in an essay employing the inductive method.

## Deductive method

A deductive argument derives one conclusion from another; the evidence tends to be other, related conclusions. You would be arguing deductively if you stated (1) that exercise is generally beneficial, (2) that walking is a form of exercise, and (3) that walking is therefore beneficial. To be convincing and logically sound, the statements or premises on which the argument is based must be true. If you said (1) that all colleges are excellent, (2) that Osmosis State is a college, and (3) that Osmosis State is therefore excellent, your argument would be faulty. Obviously you cannot argue forcefully that Osmosis State is excellent because *all* colleges are excellent when it is evident that they are not. In the deductive method, if the truth of the premises on which the conclusion is based is not self-evident, the argument collapses.

Both inductive and deductive argument must stand up to questioning. Of an inductive argument one asks whether the facts are true, whether the exceptions have been noted, whether the selection of materials is representative, whether the conclusions are truly and accurately drawn from the data, and whether the conclusions are stated precisely or exaggerated. Of a deductive argument one asks whether the given principle is impartial truth or mere personal opinion, whether it is applied to materials relevantly, whether the conclusion is accurate according to the principle, and whether exceptions have been noted.

With good motives and bad, with honesty and with deceit, different thinkers reach different conclusions derived from the same data or from the same principles. Learn to discriminate between sound and illogical reasoning. Indeed, when you write, you must constantly question your own reasoning. Watch for errors in thinking—called **logical fallacies.**

## **39a** Use only reliable sources.

The first rule of accuracy in writing is to be sure of your sources. If they are shaky, so is your argument. Outdated information is not reliable. Publications that indulge for profit in rumor and cheap sensationalism are not dependable. Nonspecialists usually are not as fully informed as specialists. People whose motives are suspect make weak witnesses. Cast a wary eye on any source that is not objective, factual, thoughtful, knowledgeable, respected, and carefully researched.

UNRELIABLE

> According to a recent issue of *The National Inquisitor,* UFO's definitely come from another planet.

UNDEPENDABLE

> Vitamins will prevent most diseases, claims the President of the VitaVitaVim Company.

SUSPECT

> Another blow against the priesthood may be the testimony of Sydney Roll, who has recently been advanced one million dollars to write a book.

## **39b** Use accurate information. Check the facts.

Facts can be demonstrated. They form the bases of judgments. Distinguish carefully between the facts and the judgments derived from them, and then explain how one comes from the other.

Errors, whether of fact, ignorance, or dishonesty, make the reader suspicious and lead to distrust. An otherwise compelling argument crumbles when just one or two facts are shown to be wrong. The following statements contain factual errors.

MISINFORMATION PRESENTED AS FACT

> Columbus was indisputably the first European to step on the North American continent.

FACT

> Some historians assert that other Europeans, especially Vikings, came to America before Columbus. Some historians believe Columbus never set foot on the North American mainland.

MISINFORMATION PRESENTED AS FACT

> Hot water will freeze faster than cold water when placed in a freezer.

FACT

> Physicists state that cold water freezes faster. (However, water that has been heated and has cooled will freeze more quickly than water that has not been heated.)

## **39c**  Avoid sweeping generalizations. Allow for exceptions.

Do not make statements about *everyone* or *everything* or *all* when you should refer to *some* or *part*. Do not be too inclusive. Such writing can be naive or deceitful. Do not convey wrong information about exceptions to a general principle. Be aware of information and opinions that seem to refute or qualify your generalizations. Deal with them fully and honestly. You can actually strengthen your argument by taking opposing opinions and exceptions into consideration.

SWEEPING GENERALIZATION

> Poor people do not got fair trials in courts

MORE CONVINCING

> Poor people often do not get fair trials in courts, in many cases because they cannot hire good attorneys.

SWEEPING GENERALIZATION

> Russian athletes are the best in the world.

MORE CONVINCING

> Russian athletes have enjoyed an extraordinary success in world competition.

CAUTION:   You cannot justify sweeping generalizations by adding a phrase such as "in my opinion." Overstatement to make a point may irritate, arouse doubt, or create disbelief. Moderation (or even understatement) convinces where brashness and arrogance alienate. Generalizations about nationalities and races are often pernicious.

## 39d Do not exaggerate to make a point.

Your credibility as a writer and thinker is damaged not only by factual errors (see **39b**) and by sweeping generalizations (see **39c**) but also by a milder and more subtle form of misrepresentation: exaggeration. Exaggeration is so common in everyday life—advertising, politics, even education—that it may seem a natural way to make a point. In fact, unless you become sensitive to the temptation of exaggeration, you may be unaware that you are indulging in it. The importance of a point may appear to warrant your magnifying beyond the facts. This is another way of saying that the end (the objective) justifies the means, a false and dangerous assumption. When you exaggerate, your argument always suffers.

## 39e Do not reason in a circle.

Reasoning in a circle, often called **begging the question**, begins with an assumption—an idea that requires proof—and then asserts that principle without ever offering proof. It states the same thing twice.

> Universal education is necessary because everyone ought to have an education.
>
> Some brokers are ambitious simply because they wish to succeed.

## 39f Avoid false comparisons.

Comparisons are highly useful in a convincing argument. Experienced writers know that a brief but fresh analogy can sometimes be more persuasive than a much longer direct explanation.

Deceptive arguments, however, sometimes try to profit from the effectiveness of analogies by making false comparisons—that is, by comparing two things that have only surface similarities.

FALSE ANALOGY

> This college is much like a Chicago slaughterhouse: neither the freshmen nor the cattle know what is in store for them, and they both get chopped up.

## **39g** Stick to the point.

An argument should be pointed. Do not attempt to be convincing by including irrelevant comments or by shifting the focus to a different issue. The **logical fallacies** discussed below stray from the point and throw up a smoke screen.

### *Red herring*

The term *red herring* originated in the practice of diverting hounds from the scent by moving a fish—a smoked (or red) herring—across their path. The following argument in favor of building a highway tries to win a point by bringing up another subject—crime—that is a separate issue, a red herring.

> Certain neighbors in our city are currently arguing against the construction of a proposed expressway. This highway should be built. Are there not more basic issues, such as crime, that these highway opponents should be addressing?

### *Argument ad hominem*

The Latin term *ad hominem* (literally, "to the man") refers to the fallacy of shifting from the real matter in question to a personal attack, diverting attention from the issue to one's opponent.

> Vote against Senator Wong's bill to provide subsistence to the poor. He has little true interest in the needy; he is a millionaire who spends much of his time on his plush yacht with his rich friends.

**39h** Do not substitute appeals to egotism or prejudice for appeals to reason.

Appeals to egotism or prejudice ignore sound reasoning and attempt to convince through other means.

### Name-calling

Simply categorizing what one does not like and giving it a name or label *(fascism, communism, racism, elitism, cultism,* and so forth*)* does not make a sound argument but appeals to preconceived notions.

> Any investigation into the private life of an elected official is nothing less than blatant McCarthyism.

### Flattery

Trying to persuade through praise rather than reason is flattery. The political candidate who tells the people that he knows they will vote for him because of their good character is indulging in this emotional appeal.

### Snob appeal

Snob appeal finds its power in the human desire to be a part of the in-crowd. It asserts that one should adopt a certain view because all the better people do.

> The best educated and most progressive readers today will no doubt agree with me that British novels are far superior to any others.

### Mass appeal

Related to snob appeal, mass appeal is the **bandwagon** approach to persuasion. It attempts to enlist support of a certain view by asserting that everyone else is supporting it.

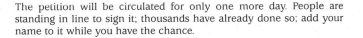
The petition will be circulated for only one more day. People are standing in line to sign it; thousands have already done so; add your name to it while you have the chance.

## 39i Draw accurate conclusions about cause and effect.

Exact causes of conditions and events are often difficult to determine, just as the effects of an action or a circumstance are. Avoid the two logical fallacies discussed below.

### Post hoc, ergo propter hoc

The Latin phrase *post hoc, ergo propter hoc* means "after this, therefore because of this." However, a condition that precedes another is not necessarily the cause of it.

> Since Beethoven's greatest music was composed after he lost his hearing, we can assume that deafness was the key to his greatest achievement.

### Non sequitur

*Non sequitur* is Latin for "it does not follow." It does not necessarily follow that one condition in a relationship is logically the effect of another.

> Good drivers make good parents.

## 39j Avoid the *either . . . or* fallacy (also called false dilemma).

Allow for adequate possibilities. When more than two choices exist, do not illogically assert that there are only two.

Either coauthors must do all their writing together, or they will not be able to write their book.

Children either do very well in school, or they remain illiterate.

## 39k Avoid illogicalities created by careless wording.

Because of a careless choice of words, what was clear in your mind can be confusing and illogical when it is expressed.

Frequently confusion and unintentional humor result when a writer who may be thinking logically does not compose logically:

When I was barely seventeen, I became engaged. My father was happy and sad at the same time. My mother was just the opposite. [How can one be the "opposite" of "happy and sad at the same time"? What the writer probably means is that her father had mixed feelings about her engagement but her mother was strongly opposed to or strongly approved of it.]

Go over your writing carefully to be sure that you are stating your thoughts logically.

## ■ Exercise 1

*Label and describe the inaccuracies and illogicalities that are in brackets.*

### The March of Progress

Ensuring progress is a lot like making pickles.
It is pretty sour business sometimes, but the results
are juicy. On the current American scene are many who
are against cutting the forests for timber, against

using animals for food and animal pelts for garments, against the use of atomic energy, and so on. Those who are against these things are against progress. Most of them are misfits with psychological problems who are trying to get attention.

We cannot turn back the clock. As the population of the world expands, so must our ideas and our technology. It simply is not wise to cancel the building of a great and needed dam to protect a minute fish like the snail darter nor to damage our timber industry and put thousands of workers out of jobs to protect the spotted owl. Why fret about saving the spotted owl when we have crime in our streets? From the early days of this great country, all reasonable people have favored the march of progress. Our first three presidents, Washington, Lincoln, and Jefferson, were outspoken advocates of progress.

A friend of mine who is employed by the U.S. Department of Agriculture states that insecticides are totally harmless to the public. Yet the loud cry goes out daily that we must ban these useful chemicals because of their threat to health. The fact that pesticides are safe argues strongly for our continued use of them to protect the crops that feed our multitudes. If those who oppose pesticides could have lived in the days when the boll weevil practically wiped out this nation's cotton crop, they might take a different view of our using chemicals. People are living longer now than they did before we used pesticides on crops; therefore, these chemicals could not possibly be a national health hazard. It is not the chemicals that we should worry about but those who demonstrate against them. After such a group brought pressure to ban pesticides in New Jersey, eight small farmers went into bankruptcy, and the nation was deprived of their produce.

Our choice is clear: we should either make up our minds to move forward in the march of progress, ignoring all of the clamor for saving or preserving this or that, or give in to these demands and give up any idea of a better way of life for ourselves and our children. I am sure that those who read this are bright enough to realize that only in the first alternative is there survival and accomplishment.

# Paragraphs

259

# **40**   Writing Paragraphs   ℱ

Paragraphs civilize writing. Without them an essay is a wilderness of sentences in which it is easy to get lost. Essays without paragraph divisions are larger versions of fused sentences, which can be frustrating or confusing. **A paragraph is a series of sentences, usually set off by indentation, grouped together for a specific reason.**

Paragraphing and thinking are inseparable. When you begin to develop initial ideas for a paper (see **41a, b, c**), you might write a page or two without any paragraph separations. Then you should sort your ideas into paragraphs—as if you were sorting your laundry into different piles. The two basic kinds of paragraphs are **function paragraphs** and **topic-sentence paragraphs.**

## Function paragraphs

Some paragraphs perform certain functions for the essay as a whole; they act as signposts in the wilderness. **Function paragraphs** include **introductions, conclusions,** and **transitions.** Writers also use function paragraphs to indicate different speakers in passages of conversation or **dialogue** (see p. 159).

### *Introductions*

The opening paragraph is often the hardest part of a paper to write. It must capture attention, introduce the subject, and state the purpose or thesis of the essay. If the first paragraph of an essay is dull, mechanical, or unclear, it may discourage further reading.

Effective ways to begin an essay include using a startling statistic or fact, telling a story or anecdote, and beginning with a generalization and moving into a specific thesis (**funnel introduction**).

In writing an opening paragraph, avoid plunging into an argument abruptly, apologizing or discussing the process of writing, and (in the funnel) using broad generalizations.

The following introduction to a paper that analyzes a magazine is not arresting:

> When I first read this magazine, I was completely bewildered, but then I started to see how it got across the idea of patriotism. After I looked at the articles, advertising, and artwork, I began to see how much the war was on everyone's mind. *The Woman's Home Companion* from 1943 is very interesting to read.

The paragraph initially establishes no context (which magazine?), then says too much about content and makeup, and wanders without a clear thesis. Below is a better way to begin the same essay:

> Fifteen cents was the cost of *The Woman's Home Companion* in 1943. In articles, advertisements, and photos, this inexpensive but colorful magazine painted a vivid picture of the people of that era. As World War II raged, Americans banded together to support their country. Consequently, in bold advertisements, many companies challenged readers to fulfill their patriotic duties.

Opening with an interesting statistic, this introduction immediately establishes a context by providing the name and date of the magazine and then narrows down to its thesis: the use of patriotism in advertisements.

### Conclusions

The concluding paragraph of a paper should be forceful and climactic. It should not repeat in detail the information presented previously. Instead, it should briefly restate or refer to the thesis and make a final assessment of the importance and originality of the paper's argument. The conclusion should attempt to answer the question "So what?"

## Transitions

The transitional paragraph usually occurs in longer essays to indicate that the author has finished discussing one major point and is now moving to another. Transitional paragraphs can be relatively brief, as the following example suggests:

> Achebe thus demonstrates how the Igbo women exercise a great deal of control in the family unit even though they have little political power in the tribe as a whole. The Ibgo men, on the other hand, rule the tribe but lack power in other areas of life.

## Topic-sentence paragraphs

The basic building block of the essay is the **topic-sentence paragraph,** a group of sentences developing one idea. What the thesis statement is to an entire paper (see p. 288), the **topic sentence** is to a paragraph. It provides a clear, crisp statement of the central thought of the paragraph, and the rest of the paragraph serves to develop, explain, and demonstrate that main idea. The following paragraph illustrates this pattern.

*topic sentence*

*body of paragraph*

Young children have no sense of wonder. They bewilder well, but few things surprise them. All of it is new to young children, after all, and equally gratuitous. Their parents pause at the unnecessary beauty of an ice storm coating the trees; the children look for something to throw. The children who tape colorful fall leaves to the schoolroom windows and walls are humoring the teacher. The busy teacher halts on her walk to school and stoops to pick up fine bright leaves "to show the children"—but it is she, now in her sixties, who is increasingly stunned by the leaves, their brightness all so much trash that litters the gutter.

ANNIE DILLARD
*An American Childhood*

**40a** Express the central thought succinctly in a topic sentence that is effectively placed. *ts*

Topic sentences most often come at the beginning of paragraphs to provide a focus for what follows.

*topic*
*sentence*

*body of*
*paragraph*

> While we waste our time fighting over ideological conformity in the scholarly world, horrible writing remains a far more important problem. For all their differences, most right-wing scholars and most left-wing scholars share a common allegiance to a cult of obscurity. Left, right, and center all hide behind the idea that unintelligible prose indicates a sophisticated mind. The politically correct and the politically incorrect come together in the violence they commit against the English language.
>
> PATRICIA NELSON LIMERICK
> "Dancing with Professors"

However, effective topic sentences can appear anywhere in a paragraph. For example, if you want to create a sense of suspense before you introduce the main subject, the topic sentence may not appear until the second or third sentence of the paragraph, as in the following example.

> Biologists may be accustomed to working on obscure beasts—ugly horseshoe crabs, prickly sea urchins, slimy algae. But anthropologist Grover Krantz of Washington State University has a problem of a different sort. He is studying, or trying to study, a creature no scientist has ever seen: the mysterious Sasquatch, or Bigfoot. [The rest of the paragraph discusses the nature of the research.]
>
> "Tracking the Sasquatch"
> *Newsweek*

When you have several examples that you intend to use to prove a point, it is sometimes effective to present the evidence first and to end the paragraph with your central idea. In the following paragraph, the topic sentence comes at the end:

Farmers no longer have cows, pigs, chickens, or other animals on their farms; according to the U.S. Department of Agriculture, farmers have "grain-consuming animal units" (which, according to the Tax Reform Act of 1986, are kept in "single-purpose agricultural structures," not pig pens and chicken coops). Attentive observers of the English language also learned recently that the multibillion dollar market crash of 1987 was simply a "fourth quarter equity retreat"; that airplanes do not crash—they just have "uncontrolled contact with the ground"; that janitors are really "environmental technicians"; that it was a "diagnostic misadventure of a high magnitude" which caused the death of a patient in a Philadelphia hospital, not medical malpractice; and that President Reagan was not really unconscious while he underwent minor surgery; he was just in a "non-decisionmaking form." In other words, doublespeak continues to spread as the official language of public discourse.

WILLIAM LUTZ
"The World of Doublespeak"

By practicing the various placements of topic sentences, you can increase your skills as an alert writer who does not compose paragraphs automatically in one monotonous pattern.

■ **Exercise 1**

*Underline the topic sentences in the following paragraphs and be prepared to discuss the effectiveness of their placement.*

1. The Polish Jewish community was entirely regulated by its rabbis in all religious, moral, and cultural matters. Its school system began with the child in elementary school and continued on up to academies of learning where masters and disciples spent their lives with the Talmud and its commentaries. In 1721 twelve years of such higher learning were required in Lithuania after marriage before one could become a teacher of the law and a judge. Jewish presses published prayer books, Bibles, the Talmud, books of Talmudic legends for light readings, and a translation of the Pentateuch into Yiddish with a

commentary woven of tales and moral teachings that was enormously appealing to women. Rabbinic civilization reached its zenith with the Jewish world of Poland and Lithuania.

CHAIM POTOK
*Wanderings*

2. The weeks until graduation were filled with heady activities. A group of small children were to be presented in a play about buttercups and daisies and bunny rabbits. They could be heard throughout the building practicing their hops and their little songs that sounded like silver bells. The older girls (non-graduates, of course) were assigned the task of making refreshments for the night's festivities. A tangy scent of ginger, cinnamon, nutmeg and chocolate wafted around the home economics building as the budding cooks made samples for themselves and their teachers.

MAYA ANGELOU
*I Know Why the Caged Bird Sings*

■ **Exercise 2**

*Compose two paragraphs, one with the topic sentence at the beginning and the other with the topic sentence in another position. Be prepared to justify the placement of the topic sentences.*

**40b** Unify a paragraph by relating each sentence to the central idea.

Each sentence in an effective topic-sentence paragraph is directly related to the main point. Do not make your readers strain to find connections. Even slightly irrelevant material can throw the entire paragraph out of focus and lead to confusion about the direction of the argument. When you shape your freewriting or brainstorming (see pp. 284–287) into paragraphs, you probably will find extraneous material that you should omit.

The following paragraph opens with a good topic sentence but then, after the second sentence, begins to stray from the main point.

> Although heroes can provide good role models, hero worship can also have negative effects. Heroes can encourage us to strive for higher goals and are fun to have. *Lots of people read about their heroes in the newspaper or watch them on television. But sometimes people spend too much time watching television, which has lots of violence.* I have watched the runner Mary Decker on television a lot and have been inspired to become a better athlete. But trying to copy the achievements of a hero can also lead to frustration. A woman may train as hard as she can and never be as fast as Decker, which could lead to great disappointment.

What starts out as a consideration of the pros and cons of heroes veers off into a comment on television before returning to the idea of heroes. Below is the paragraph revised for better unity.

> Although heroes can provide good role models, hero worship can also have negative effects. Heroes can encourage us to strive for higher goals. For example, Mary Decker's successes in running have inspired me to become a better athlete. But trying to copy the achievements of a hero can also lead to frustration. A woman may train as hard as she can and never be as fast as Decker, which could lead to great disappointment.

Often it is not so easy to make a good paragraph from a flawed one. The paragraph below starts off on one topic but halfway through it switches to another.

> Mountain bicycles are a practical mode of transportation in the city. Their wide, knobby tires easily bounce over potholes, rough patches, and railway ties. The urban mountain-bicycle rider also has superior visibility since he or she sits in an upright position rather than leaning over the handlebars. Those *Tour de France* bicycle racers

must get tired backs, stretched out over their handlebars all day long. These professionals are tremendously strong and have amazing endurance, for this race lasts several days and takes them over numerous mountain passes.

The author must decide what the paragraph will be about: the practical aspects of mountain bicycles or the qualities of a *Tour de France* racer. Nonunified paragraphs often emerge when a writer is not sure what he or she wants to say.

■ Exercise 3

*Study the paragraph below, and delete all extraneous material.*

Many American couples today have chosen not to follow the traditional way of being parents, in which the mother serves as the primary caregiver. Both mother and father take family leave when the baby is born, and both feed and change the newborn. During its first year, a baby is very demanding. Some children suffer from colic, which makes them cry for long hours and refuse to go to sleep. The first year can be a tiring time for parents. When the child is older, coparents both participate in the daily routines of caregiving: cooking, feeding, bathing, entertaining, and disciplining. Consequently, the child bonds equally with both the mother and the father.

# 40c Develop paragraphs adequately.

Good paragraphs provide enough information to demonstrate the ideas in their topic sentences. To develop a paragraph, provide specific details, examples, or explanations that support the generalization given in the topic sentence. Skilled writers move back and forth between generalizations and the specific details that flesh out those generalizations.

Starting a new paragraph after every two or three sentences, without regard to development, is nearly as distracting as having no paragraph divisions at all. Underdeveloped paragraphs are like undernourished people: they are weak, unable to carry out their assigned tasks. Make sure your paragraph divisions come with new units of thought and that you have fleshed out an idea appropriately before going on to another idea.

The following paragraphs on George Washington are ineffectively developed:

> In many ways, the version of George Washington we grew up on is accurate. He was an honest and personally unambitious politician.
>
> But history teachers rarely tell us about the complex character of the man.
>
> Most historians feel that he was not a great soldier, yet he was the luckiest man alive.
>
> He even survived the attacks of his mother. When he was president, she went so far as to state publicly that he was starving and neglecting her, neither of which was true.

Though interesting, these paragraphs are too underdeveloped to do their jobs. They are just strong enough to state the points. Contrast them with the paragraphs below:

> In many ways, the version of George Washington we grew up on is accurate. He was an honest and personally unambitious politician, a devout patriot and a fearless soldier. Through circumstance, he became the lodestar for the swift-sailing Revolution. Even his enemies conceded that national success would have been impossible without him.
>
> But history teachers rarely tell us about the complex character of the man. Often moody and bleak, he did not like to be touched. When he became president, he required so much ritual and formality that it caused one observer to quip: "I fear we may have exchanged George the Third for George the First." Though he drank hard with the enlisted men, he was a tough disciplinarian, describing himself as a man "who always walked on a straight line."

Most historians feel that he was not a great soldier, yet he was the luckiest man alive. He had two horses shot out from under him; felt bullets rip through his clothes and hat; and survived attacks by Indians, French and English troops, hard winters, cunning politicial opponents, and smallpox.

He even survived the attacks of his mother. When he was president, she went so far as to state publicly that he was starving and neglecting her, neither of which was true. Then, to twist the emotional knife, she tried to persuade the new government to pass a law that future presidents not be allowed to neglect their mothers.

Adapted from DIANE ACKERMAN
"The Real George Washington"

In certain kinds of writing, short paragraphs are acceptable and expected. Newspapers, for example, use brief paragraphs because their column width is narrow, and their stories are short. Sometimes, in an expository essay, it is effective to insert a short paragraph amid longer ones for emphasis and variety. The length of a writing assignment may influence the lengths of paragraphs. A paper of one thousand words provides more room to develop full paragraphs than a short assignment does.

■ **Exercise 4**

*Seven broad subjects for paragraphs are listed below. Choose the two that you like best, and select one aspect of each topic. On each subject, first write a skimpy paragraph. Then write a paragraph of 125 to 175 words with fuller development of the subject.*

1. diets
2. E-mail
3. television soap operas
4. condominiums
5. the volunteer army
6. college sports
7. national healthcare

## **40d**   Trim, tighten, or divide sprawling paragraphs.

A sprawling paragraph is the opposite of an underdeveloped one: it needs trimming. As you write, keep alert to the length of your paragraphs. One that extends over several pages is obviously too long. To reduce excessive length, you may need to reduce the scope of the topic sentence or divide the material into smaller units. Sometimes you can produce a better paragraph by simply discarding material. For example, all your details may be pertinent and interesting, but you may not need ten examples to illustrate your point; four or five may do it more efficiently.

## **40e**   Develop paragraphs by appropriate methods.

To flesh out the main idea being presented in a paragraph, you can follow a number of different traditional **patterns of development.** Often the topic sentence of the paragraph will suggest a logical method of development. Although one pattern might dominate a paragraph, most paragraphs use a combination of patterns. For example, a paragraph of comparison and contrast might use definitions, and an explanation of process might employ narration. Use those patterns that help deliver your ideas most effectively.

### *Narration*

Narrative writing unfolds events over time and sometimes across space. In many ways, this is the simplest pattern because we tend to think in chronological order. Narratives usually begin with the first event and end with the last. In some instances, however, it is effective to describe a later event first and then go back to the beginning (employing a flashback).

The following two paragraphs describe events in chronological order:

The taxi was almost upon her before she noticed it. As it slowed down for the gate, she identified it by its lights, and stepped on to the roadway. It slid to a halt in the gravel, beside her; she stepped in before the driver could come to assist her, slammed the door, and said: "Wooford's Drugstore, on Manchester Street."

He was waiting in front of the drugstore. Almost a block before they got there, she could see him pacing back and forth with his almost prinking, elastic step, his shoulders in the slight slouch which he affected, a hat crushed on his head, and his old raincoat swinging free from the shoulders with the loose belt trailing at both ends. He looked around when the cab pulled to a stop at the corner, and strolled over to her, and watched critically while she paid off the driver. As she turned to face him, he said: "You tipped him far too much."

ROBERT PENN WARREN
*At Heaven's Gate*

In the above paragraphs, the writer describes the events as they happened.

## Description

Descriptive writing appeals to the five senses: sight, hearing, smell, taste, and touch. Descriptive paragraphs can follow a chronological, spatial, least-to-most-important, or most-to-least-important pattern. In describing a scene or a person, move from detail to detail in a logical order.

In the following paragraph, the reader's eye is guided around a house under construction. The author also arranges the description in climactic order, ending the scene with the most important person present.

Inside the house, carpenters, electricians, and plumbers were busily at work. Plywood covered the floor, and everything was shrouded in a thin film of white dust. A plasterer on a ladder was touching up a molding; an older man in overalls stood at a table saw, cutting strips of wood flooring; two other workers—electricians, to judge from the tool belts slung around their waists—were conferring over a set of blueprints spread out on a makeshift trestle table. A well-dressed woman—obviously the owner—surveyed this activity with what appeared to be a mixture of pride and trepidation.

WITOLD RYBCZYNSKI
*Looking Around: A Journey Through Architecture*

NOTE:  Paragraphs of **narration** and **description** sometimes do not have conventional topic sentences.

### Example

Many topic sentences state generalizations that gain force with evidence and illustration. Proof can be provided by an extended example or by several short examples. Well-chosen examples can add interest and vividness as well as proof, as can be seen in the following paragraph:

> Ours was once a forested planet. The rocky hillsides of Greece were covered with trees. Syria was known for its forests, not its deserts. Lebanon had vast cedar forests, from which the navies of Phoenicia, Persia, and Macedonia took their ship timber, and which provided the wood that Solomon used to build the temple at Jerusalem. Oak and beech forests dominated the landscapes of England and Ireland. In Germany and Sweden, bears and wolves roamed through wild forests where manicured tree farms now stand. Columbus saw the moonscape that we call Haiti "filled with trees of a thousand kinds." Exploring the east coast of North America in 1524, Verrazano wrote of "a land full of the largest forests . . . with as much beauty and delectable appearance as it would be possible to express."
>
> CATHERINE CAUFIELD
> "The Ancient Forest"

### Definition

Whether you are using a new word or a new meaning for an old one, a definition explains a concept. It avoids the problems that arise when two persons use the same term for different things. Effective definitions use specifics or examples to demonstrate abstractions.

A paragraph of definition may be much more elaborate than a definition given in a dictionary.

> "Whistleblower" is a recent label for those who make revelations meant to call attention to negligence, abuses, or dangers that threaten the public interest. They sound an alarm based on their expertise or inside knowledge, often from within the very organization in which they work. With as much resonance as they can muster, they strive to breach secrecy, or else arouse an apathetic public to dangers everyone knows about but does not fully acknowledge.
>
> SISSELA BOK
> *Secrets: On the Ethics of Concealment and Revelation*

## Analogy

An analogy is a figurative comparison; it explains one thing in terms of another. Analogies are most effective when you show a resemblance between two things not ordinarily compared to each other. Here, Patricia Limerick explains writing in terms of carpentry:

> A carpenter . . . makes a door for a cabinet. If the door does not hang straight, the carpenter does not say, "I will *not* change that door; it is an expression of my individuality; who cares if it will not close?" Instead, the carpenter removes the door and works on it until it fits. That attitude, applied to writing, could be our salvation. If we thought more like carpenters, academic writers could find a route out of the trap of ego and vanity. Escaped from that trap, we could simply work on successive drafts until what we have to say is clear.
>
> PATRICIA NELSON LIMERICK
> "Dancing with Professors"

## Comparison and contrast

Comparisons show similarities; contrasts present differences. The ratio of comparison and contrast can vary greatly from one paragraph to another. You might write, for example, that two things have five likenesses but only one or two differences. Or the reverse could be true.

The following paragraph on Thomas Jefferson and Abraham Lincoln begins by briefly contrasting the two men and then moves on to point out three similarities in their early lives.

> As one would expect, the formative years of Jefferson and Lincoln represent a study in contrasts, for the two men began life at opposite ends of the social and economic spectrum. There are, however, some intriguing parallels. Both men suffered the devastating loss of a parent at an early age. Jefferson's father, an able and active man to whom his son was deeply devoted, died when his son was fourteen, and Thomas was left to the care of his mother. His adolescent misogyny and his subsequent glacial silence on the subject of his mother strongly suggest that their relationship was strained. Conversely, Lincoln suffered the loss of his mother at the age of nine, and while he adored his father's second wife, he seems to have grown increasingly unable to regard his father with affection or perhaps even respect. Both Jefferson and Lincoln had the painful misfortune to experience in their youth the death of a favorite sister. And both were marked for distinction early by being elected to their respective legislatures at the age of twenty-five.
>
> DOUGLAS L. WILSON
> "What Jefferson and Lincoln Read"

NOTE: Sometimes entire paragraphs may consist of either comparisons or contrasts rather than of both.

Paragraphs using comparison and contrast can follow an **alternating structure,** which moves back and forth between the subjects being discussed, or a **block structure,** which discusses one subject completely and then moves to the next. In the paragraph below, the author uses an alternating structure to contrast Jefferson and Lincoln.

*Jefferson*       In the matter of education the contrasts are equally great. Jefferson received a superb education, even by the standards of his class. It included formal schooling from the

| | |
|---|---|
| *Lincoln* | age of five, expert instruction in classical languages, two years of college, and a legal apprenticeship. Along the way he had the benefit of conspicuously learned men as his teachers. . . . Lincoln had almost no formal education. Growing up with nearly illiterate parents and in an atmosphere that had, as he wrote, "nothing to excite ambition for education," Lincoln was essentially self-taught. The backwoods schools he attended very sporadically were conducted by teachers with meager |
| *Jefferson* | qualifications. . . . Jefferson read Latin from an early age and, after mastering classical languages and French, was able to |
| *Lincoln* | teach himself Italian; Lincoln at about the same age was teaching himself grammar in order to be able to speak and write standard English. |

If the author had used the block structure for the same paragraph, it would read as follows.

| | |
|---|---|
| *Jefferson* | In the matter of education the contrasts are equally great. Jefferson received a superb education, even by the standards of his class. It included formal schooling from the age of five, expert instruction in classical languages, two years of college, and a legal apprenticeship. Along the way he had the benefit of conspicuously learned men as his teachers. He read Latin from an early age and, after mastering classical languages and French, was able to teach himself Italian. |
| *Lincoln* | Lincoln had almost no formal education. Growing up in an atmosphere that had, as he wrote, "nothing to excite ambition for education," Lincoln was essentially self-taught. The backwoods schools he attended very sporadically were conducted by teachers with meager qualifications. Lincoln taught himself grammar in order to be able to speak and write standard English. |

Adapted from DOUGLAS L. WILSON
"What Jefferson and Lincoln Read"

## *Cause and effect*

Generally a paragraph of this kind states a condition or effect and then proceeds by listing and explaining the causes. However, the first sentences may list a cause or causes and then conclude with the consequence, the effect. In either method, the paragraph usually begins with a phenomenon that is generally known and then moves on to the unknowns.

The following paragraph begins with an effect and proceeds to examine the causes:

> This close-knit fabric [of the city] was blown apart by the automobile, and by the postwar middle-class exodus to suburbia which the mass-ownership of automobiles made possible. The automobile itself was not to blame for this development, nor was the desire for suburban living, which is obviously a genuine aspiration of many Americans. The fault lay in our failure, right up to the present time, to fashion new policies to minimize the disruptive effects of the automobile revolution. We have failed not only to tame the automobile itself, but to overhaul a property-tax system that tends to foster automotive-age sprawl, and to institute coordinated planning in the politically fragmented suburbs that have caught the brunt of the postwar building boom.
>
> EDMUND K. FALTERMAYER
> *Redoing America*

### Classification and division

Classifying and dividing are related to comparing and contrasting. **Classification** stresses similarities or common denominators; **division** (or **analysis**, as it is sometimes called) stresses differences. To show how two or more things are related, despite their differences, and then to label that grouping is to classify. To show how despite being part of the same group or class, two or more things are distinctive is to divide (or analyze).

The following paragraph classifies several famous people, assigning each to either of two categories: extroverts and introverts.

> Most people can be classified as either extroverts or introverts. William Shakespeare, Franz Schubert, George Armstrong Custer, and Charles Dickens form a disparate group indeed, but they had one important characteristic in common: the communal urge. They

sought to find their identity in relation to other people and thus can be termed extroverts. On the other hand, Emily Dickinson, Joseph Stalin, Herbert Hoover, and Geronimo exhibited strong proclivities toward inwardness and constantly had to struggle with the urge to exclude others from their activities and interests. Different as they were, they shared the distinction of being introverts.

Contrast the paragraph above, which illustrates a method of classification, with the following paragraph, in which the writer breaks down one of the categories—that of introverts—into component parts and analyzes them, thus developing the paragraph by the method of division.

Although introverts share broad characteristics, they differ greatly among themselves. Aggressive introverts, such as Joseph Stalin, find their tendency toward privacy disturbing and attempt to deal with it, indeed to obliterate it, by forcing themselves into positions of power and authority. Creative introverts, like Emily Dickinson, tend to possess greater self-knowledge and to accept themselves for what they are. They become adjusted to themselves (though never to society), and that leads to a high potential for creativity.

### Process

Several methods can be used in describing a process. Most processes are given in chronological order, step by step. The simplest kind of process perhaps is the type used in a recipe, usually written in the second person or imperative mood. The necessity here is to get the steps in order and to state each step very clearly. This process is instruction; it tells *how to do* something.

Another kind is the exposition of *how something works* (a clock, the human nervous system). Here the writer should avoid technical terms and intricate or incomprehensible steps. This kind of process is explanation.

Still another kind of process, usually written in the past tense, tells *how something happened* (how oil was formed in the earth, how a celebration or a riot began). Usually a paragraph of this type is written in the third person. It is designed to reveal how something developed (once in a single time period or on separate occasions). Its purpose is to explain and instruct.

The following paragraph explains one theory about the process that formed the moon. Distinct steps are necessary for explanation here just as they are for instruction in a paragraph that tells how to do something.

> There were tides in the new earth long before there was an ocean. In response to the pull of the sun the molten liquids of the earth's whole surface rose in tides that rolled unhindered around the globe and only gradually slackened and diminished as the earthly shell cooled, congealed, and hardened. Those who believe that the moon is a child of earth say that during an early stage of the earth's development something happened that caused this rolling, viscid tide to gather speed and momentum and to rise to unimaginable heights. Apparently the force that created these greatest tides the earth has ever known was the force of resonance, for at this time the period of the solar tides had come to approach, then equal, the period of the free oscillation of the liquid earth. And so every sun tide was given increased momentum by the push of the earth's oscillation, and each of the twice-daily tides was larger than the one before it. Physicists have calculated that, after 500 years of such monstrous, steadily increasing tides, those on the side toward the sun became too high for stability, and a great wave was torn away and hurled into space. But immediately, of course, the newly created satellite became subject to physical laws that sent it spinning in an orbit of its own about the earth. This is what we call the moon.
>
> RACHEL CARSON
> *The Sea Around Us*

## ■ Exercise 5

*Identify which method or methods of development the following topic sentences suggest.*

1. The car designs this year are unusually innovative.

2. Advertising often influences consumers when they decide what kind of clothing to buy.

3. Changing a baby's diaper is simple.

4. During the first week of college, I felt as if I were drowning.

5. The messages people record on their answering machines can be functional, comical, or simply bizarre.

6. Life in the Midwest is different from life on the West Coast.

## **40f**   Use transitional devices to show the relationships between the parts of your writing.   *tr*

Transitional devices are connectors and direction-givers. They connect words to other words, sentences to sentences, paragraphs to paragraphs. Writings without transitions are like a strange land with no road signs. Practiced writers assume that they should give clear signs as to where a paragraph and a paper are going.

The beginnings of paragraphs can contribute materially to clarity, coherence, and the movement of the discussion. Some writers meticulously guide readers with a connector at the beginning of almost every paragraph. H. J. Muller, for example, begins a sequence of paragraphs about science as follows.

> In this summary, science . . .
> Yet science does . . .
> Similarly, the basic interests of science . . .
> In other words, they are not . . .
> This demonstration that even the scientist . . .
> This idea will concern us . . .
> In other words, facts and figures . . .

CONNECTIVE WORDS AND EXPRESSIONS

| | | |
|---|---|---|
| but | indeed | likewise |
| and | in fact | consequently |
| however | meanwhile | first |
| moreover | afterward | next |
| furthermore | then | in brief |
| on the other hand | so | to summarize |
| nevertheless | still | to conclude |
| for example | after all | similarly |

DEMONSTRATIVE PRONOUNS

    this       that      these      those

References to demonstratives must be clear (see p. 65).

OTHER PRONOUNS

    many    each    some    others    such    either

### Repeated key words, phrases, and synonyms

Repetitions and synonyms guide the reader from sentence to sentence and paragraph to paragraph.

### Parallel structures

Repeating similar structural forms of a sentence can show how certain ideas within a paragraph are alike in content as well as structure. A sequence of sentences beginning with a noun subject or with the same kind of pronoun subject, a series of clauses beginning with *that* or *which*, a series of clauses beginning with a similar kind of subordinate conjunction (like *because*)—devices like these can achieve transition and show connection.

Excessive use of parallelism, however, is likely to be too oratorical, too dramatic. Used with restraint, parallel structures are excellent transitional devices.

The following paragraph on the subject of emotions illustrates how various transitional devices can help create direction and coherence.

Emotions are part of our genetic heritage. Fish swim, birds fly,

└── **repeated words** ──┐

and people feel. Sometimes we are happy; sometimes we are not.

**connective word: But**

But sometimes in our life we are sure to feel anger and fear, sadness

and joy, greed and guilt, lust and scorn, delight and disgust. While

we are not free to choose the emotions that arise in us, we are free

to choose how and when to express them, provided we know what

**demonstrative: That**

they are. That is the crux of the problem. Many people have been

**parallel structure**

educated out of knowing what their feelings are. When they hated,

they were told it was only dislike. When they were afraid, they were

told there was nothing to be afraid of. When they felt pain, they were

advised to be brave and smile. Many of our popular songs tell us

"pretend you are happy when you are not."

HAIM G. GINOTT
*Between Parent and Child:*
*New Solutions to Old Problems*

## General exercises

### ■ Exercise 6

*Write three paragraphs on any of the following subjects. Use a different primary method of development for each. Name the various methods you use in each paragraph.*

1. good manners
2. autumn
3. cooking
4. farms

5. wealth
6. political campaigns
7. the Berlin Wall
8. tennis

### ■ Exercise 7

*Find three good paragraphs from three different kinds of writing: from a book of nonfiction, an essay, a review, or a newspaper article. Discuss how effective paragraphs differ in different kinds of writing.*

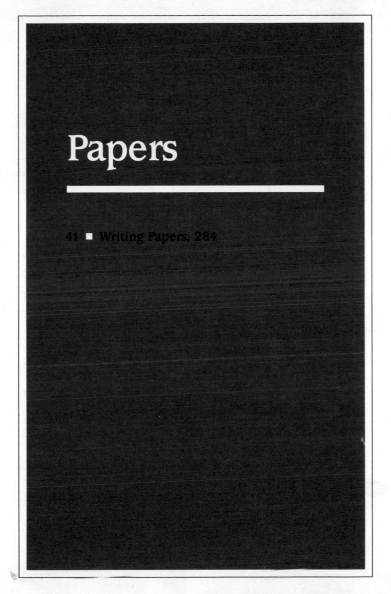

# Papers

---

# 41 Writing Papers

A good paper usually evolves through a number of stages, including preliminary thinking (brainstorming), jotting down ideas and writing in a free-flowing way (freewriting), composing a draft (or drafts), revising, and writing the final version.

## 41a Generate and develop ideas.

Whether your instructor assigns you a topic to write about or expects you to discover your own, you should begin by coming up with ideas and getting them on paper or on the screen of your computer. Some techniques proven effective in generating and developing ideas are discussed below.

### Brainstorming

Brainstorming begins with opening up the mind to sudden flurries of ideas. This mental activity is often associated with unpredictable inspiration, but with some practice you can initiate it whenever you wish. Force yourself to open your mind to all possibilities and to make free associations. As ideas occur to you, get them down rapidly on paper or on your computer screen. Later you can go over the results of your brainstorming to determine which ideas to keep and which to reject.

### Journal jotting

Many a good idea for a paper has been lost—that is, forgotten—because it was not recorded when it emerged. You cannot depend on even a sharp and reliable memory to retain fleeting ideas. Therefore, develop the habit of **journal jotting.** Some of the world's greatest writers keep a journal and have it available at all times so that they can write down thoughts that can later be developed. Unless you want to make the entries extensive, you need not do so; jot down just enough to make sure you can remember a promising idea at a later time.

### Listing

You may find it helpful to **list** various points suggested by your general subject. At this stage you do not need to be concerned with their order or final relevance. In fact, you will probably list more points than you will want to make in your paper. The purpose of listing, and of other prewriting strategies, is to get your thoughts on paper and to stimulate further thinking.

Suppose a student is interested in current advertising and has decided to write a paper on some aspect of this subject. He or she might list the following thoughts:

1. recent advances in ways of presenting products
2. doctors and attorneys who advertise
3. selling by mail
4. news commentators who also deliver commercials
5. opportunities in this field for high-paying positions
6. may be the most competitive of all fields at present
7. government control of advertising—is it a good idea?
8. for successful firms, is advertising really necessary?
9. effects of disagreeable advertisements on sales
10. artistic content of some television advertisements
11. talk shows that allow guests to plug latest motion picture or book
12. deceptiveness in methods of advertising
13. advertisements in newspapers that are made to look like news items
14. appeals for money on religious programs
15. false advertising
16. the high cost of television advertising
17. dangers of disguised advertising—brings on loss of trust
18. ethics—question of whether still important to those who control advertising

19. advertising and mind control
20. subliminal messages

The writer has randomly listed almost everything that came to mind—too much for a paper of average length. The list is a good start, however, in exploring the subject of advertising.

## *Clustering*

Relationships among several ideas are often easier to understand when you project them graphically. In **clustering** you draw a diagram rather than write. Just as graphs and maps are useful alternatives to words, this technique of invention and development can aid in revealing connections among ideas. Begin by writing down your principal subject and circling it. Then as sub-ideas come to you, jot them down, circle them, and draw connecting lines to reveal relationships that you have perceived.

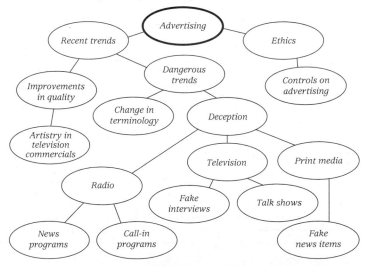

A large blackboard is ideal for this kind of exercise when you are developing a subject, working through a mental puzzle. You can erase frequently and substitute new ideas. If a blackboard is not available, you can achieve much the same effect with pencil and paper. However you do it, clustering enables you to view the subject from different perspectives and to see graphically how the composite parts are related.

## Freewriting

Some writers prefer to think a paper through in their heads before they begin writing. They think and plan; then they write and revise. Others trying to follow this method encounter difficulties; they find it hard to get started without a boost. One way to get started is by expressing in words any thoughts that come to mind—anything at all:

> This is an impossible assignment. What can I find to say about an essay that deals with personal responsibility? I heard a talk-show host going on and on about this subject the other day. Blab, blab, blab. Rights without responsibility. Whatever I think of seems so obvious. Anyway, I'm hungry and need to order a pizza or something. How can you think on an empty stomach? Some people act on an empty head. Maybe that's a good idea.

At some point, an idea emerges that may be just what you need, the boost. This activity, called **freewriting**, takes a writer wherever whim dictates. Nothing is ruled out. You put down on paper or on the computer screen whatever pops into your mind.

Freewriting usually does not produce a first draft, but it can generate ideas and reveal connections. It is a way of discovering what you want to write about as you write, a way of overcoming the inhibitions some people experience when they view a blank sheet of paper or a blank screen.

SUGGESTION: If you are worried about the possibility of wasting precious time by writing down random thoughts, set a time limit for freewriting—say, fifteen minutes. That should be enough to get you started.

## **41b**   Compose a precise thesis statement.

Ideas emerging from the preliminary processes described above are likely to be somewhat in disarray. They need to be examined closely, culled, and ordered. This stage of the writing process is necessary if the finished paper is to be coherent and sharply focused, not rambling and pointless.

At some point you must determine precisely what it is you wish to say. Each paper should have a clear center from which all the details and illustrations radiate.

When you pin down the central idea, you can usually express it in a single sentence, called the **thesis statement.** Many instructors require you to compose a specific and concise statement and to include it in your first or second paragraph to develop a good sense of direction.

A thesis statement tells in a few words what your argument is, not what your argument is *about*. Compare the three statements below (see model paper "Conduct and the Inner Self," pp. 312–315, written in response to Willard Gaylin's essay "What You See Is the Real You," pp. 295–297.

NOT

> The purpose of this paper is to agree basically with Willard Gaylin's essay "What You See Is the Real You." [A description, not a thesis statement.]

NOT

> In dealing with people, the profession of psychoanalysis often gives society the wrong impression about moral responsibility. [A thesis statement, but not a good one because it is not specific enough.]

BUT

> Studying and emphasizing "the inner man," as Willard Gaylin argues in "What You See Is the Real You," has had the unfortunate effect of relieving individuals of moral responsibility for their acts.

# 41c Organize effectively.

What to place where can be a problem—although sometimes the parts fall naturally into place. You may or may not need an outline, depending on the extent to which the development process has already unfolded in your mind. It is a good idea to set out with some sort of plan, though, even if you require only a brief one. Three of the most common kinds of outlines are described below.

### Scratch outline

This is the simplest kind of outline: a list of points to make, in order but without subdivisions. It is a quick way to organize your thoughts and to remind yourself of their order while you are writing. For brief papers, those written in class, and essay examinations, a scratch outline usually suffices. You might use the following points for a scratch outline of the paper entitled "Conduct and the Inner Self."

> Gaylin right in pointing out modern tendency to stress inner self
> importance of accepting moral responsibility for actions
> history proves this
> Gaylin in error in some of his illustrations
> soundness of his basic point

### Topic outline

The topic outline is a formal, detailed structure to help you organize your materials. Observe the following conventions:

1. Number the main topics with Roman numerals, the first subheadings with capital letters, and the next with Arabic numbers. If further subheadings are necessary, use a, b, c and (1), (2), (3).

```
I. . . . . . . . . . . . . . . . . . . . . . . . . . . . . . . . . . . . . . . . . . .
   A. . . . . . . . . . . . . . . . . . . . . . . . . . . . . . . . . . . . . . . . .
      1. . . . . . . . . . . . . . . . . . . . . . . . . . . . . . . . . . . . . . . .
         a. . . . . . . . . . . . . . . . . . . . . . . . . . . . . . . . . . . . . .
            (1) . . . . . . . . . . . . . . . . . . . . . . . . . . . . . . . . . .
            (2) . . . . . . . . . . . . . . . . . . . . . . . . . . . . . . . . . .
         b. . . . . . . . . . . . . . . . . . . . . . . . . . . . . . . . . . . . . .
      2. . . . . . . . . . . . . . . . . . . . . . . . . . . . . . . . . . . . . . . .
   B. . . . . . . . . . . . . . . . . . . . . . . . . . . . . . . . . . . . . . . . .
  II. . . . . . . . . . . . . . . . . . . . . . . . . . . . . . . . . . . . . . . . . .
```

2. Use parallel grammatical structures.
3. Write down topics, not sentences.
4. Do not place periods after the topics.
5. Punctuate as in the example that follows.
6. Check to see that your outline covers the subject fully.
7. Use specific topics and subheadings arranged in a logical, meaningful order.
8. Be sure that each level of the outline represents a division of the preceding level and has a smaller scope.
9. Avoid single subheadings. If you have a Roman numeral I, you should also have a II, and so forth.

INCORRECT

```
   I. Gaylin's essay as a corrective to some current thinking
      A. Nature of faulty thinking on individual responsibility
←───────── B.  [Another subheading is needed.]
  II. Gaylin's argument against the reality of the "inner man"
```

Following is an example of a topic outline for the paper "Conduct and the Inner Self":

Conduct and the Inner Self

```
   I. Gaylin's essay as corrective to some current thinking
      A. Nature of faulty thinking about individual responsibility
      B. How study of "inner man" contributes to faulty thinking
```

II. Gaylin's argument against the reality of the "inner man"
   A. Purpose of argument
   B. Argument from historical perspective
III. Problems in Gaylin's presentation
   A. Exaggeration
      1. Praise of a hypocrite
      2. Total denial of the good-heart principle
   B. Contradiction
IV. Validity of Gaylin's argument despite flaws

## Sentence outline

The sentence outline is an extensive form of preparation for writing a paper. More thinking goes into it than into a scratch or topic outline, but the additional effort usually enables you to keep tight control over your writing. The sentence outline follows the same conventions as the topic outline, but the entries are expressed in complete sentences. Place periods after sentences in a sentence outline.

### Conduct and the Inner Self

I. Gaylin's essay is a much-needed corrective to faulty thinking today.
   A. Currently there is a tendency to excuse those who act irresponsibly.
   B. Emphasizing the "inner man" tends to relieve individuals of moral responsibility.
II. Gaylin's argument against the reality of the "inner man" is ingenious and sound.
   A. The purpose of the argument is to stress that no aspect of our being is more important than our conduct.
   B. History has proven Gaylin correct in his insistence on personal responsibility.
III. Some aspects of Gaylin's argument present problems.
   A. He exaggerates to make a point.
      1. Praise of the old man who lived a life of hypocrisy is unconvincing.
      2. Gaylin's denial that a heart of gold can lie beneath bad behavior contradicts experience.

B.　Gaylin contradicts himself when he argues against the exis-
　　　　tence of an inner self and then speaks as if there were such
　　　　a thing.
　IV.　Though flawed, Gaylin's essay makes a valid and timely point.

# **41d**　Adapt your writing to your audience.

Writing a paper is an act of communication, not merely an exercise
or an opportunity for self-expression. Prepare yourself to reach
your audience effectively. Consider the following questions. How
much information do members of the audience have on the subject
you are dealing with? What is their average level of education? Are
they likely to have strong opinions one way or the other on the
topic? What lines of work or study are they engaged in as a rule?
Do they fall into a definite age group? Writers who ignore such
important considerations are inviting rejection. Mark Twain
learned the necessity of carefully matching writing with audience
when he composed and delivered a humorous but rather crude
series of caricatures to an audience of distinguished and highly
dignified New England writers whom he greatly admired and did
not wish to insult. The performance was received with stony
silence by the offended audience, and Mark Twain regretted his
misjudgment for life.

　　Although your instructor may ask you to write for specialized
audiences, usually papers are composed with *general readers* in
mind: educated adults who do not have highly specialized vocabularies and who are interested in a wide variety of general subjects.
To capture their attention, you must be intelligent, interesting,
rational, and clear. To keep their attention and win their respect,
you must be fair and remain sensitive to their feelings. For example, you should be constantly aware that your audience consists of
both men *and* women and avoid using generic masculine pronouns
(*he* or *his*) and terms like *man* and *mankind* (for *humankind*). Notice
that Willard Gaylin uses terms of masculine gender ("inner man")
in his thought-provoking essay on pp. 295–297 and thus runs the

risk of offending and alienating a segment of his audience. Study the model paper (pp 312–315) written as a response to Gaylin's essay, and observe how the author of the paper wisely avoids language that could be interpreted as sexist (see also **8f**).

Resist the temptation to write solely for an individual professor. Frustrated students who complain that they do not know what their professors want in a paper can often avoid this needless exercise in trial and error by concentrating on a general audience. If you compose with only your professor in mind, you may write somewhat over your head or employ obscure or flowery language.

Audience awareness helps you to become a better judge of your own writing and consequently to revise more successfully. By learning to read your writing as your audience will, you can communicate with greater clarity and grace.

■ **Exercise 1**

*Think of an experience in which you felt that you were wronged or in which you wronged someone: something involving a store, an automobile repair shop, a restaurant, an insurance company, or the like. Then write about this experience three times, with three audiences in mind: (1) close friends; (2) managers, supervisors, or owners of the business; (3) readers of a large local newspaper. All three writings may be in the form of letters.*

## **41e** Sustain an appropriate tone.

Tone derives from attitude. If you are displeased, you can show that attitude through your choice of words. If you are amused and wish to pass that attitude on to your audience, you can do so by creating a humorous tone. Enthusiasm, urgency, fear, objectivity, displeasure, playfulness, skepticism—all can be conveyed effectively, but first you have to determine precisely what attitude you wish to communicate. Trouble occurs when a writer is uncertain what tone to take and suddenly switches unthinkingly from one attitude to another, as in the following example:

> Art critics have the power to make or break young painters. Their word is especially influential among collectors and wealthy investors who deal in art. If a few of these better-known critics praise a given artist, he or she can rise almost overnight from struggling anonymity to wealthy repute. Conversely, all it takes is a negative opinion from them *shift* to doom an aspiring artist. Do these critics exercise their enormous *in* power responsibly? They care only for themselves. They are shame- *tone* less frauds who invent nonsense about art and pass it off as the truth. They are arrogant and effete impostors who deserve to be whipped.

At the beginning of the passage, emotions are under control and the tone is coolly objective. Then the tone changes abruptly as an angry outburst destroys what could otherwise be a convincing argument. If the writer had decided ahead of time to make indignation the prevailing tone of the essay, he or she could have written indignantly but thoughtfully and effectively. As it is, the writing is ineffectual because the sudden shift in tone confuses readers.

Whatever tone you choose, sustain it carefully; do not shift back and forth from one attitude to another. In general, avoid flippancy and sarcasm. Flippancy in writing is often the mark of immaturity. Sarcasm is difficult to sustain without creating hostility.

## **41f** Compose a draft before writing the final version.

A rough draft is a necessary step toward a finished paper. Writers who go directly from an outline to what they consider the final copy usually produce a less effective paper than they could otherwise. Do not worry about how you are expressing your ideas in your draft; simply get them down. Then you can make improvements.

The rough draft on pp. 297–301 emerged from an assignment in which the student was required to read a specific essay and then to respond to it in a paper of about five hundred words. First read the essay on which the assignment was based, "What You See Is the Real You" by Willard Gaylin. Then read the student's rough draft.

## What You See Is the Real You
### Willard Gaylin

It was, I believe, the distinguished Nebraska financier Father Edward J. Flanagan who professed to having "never met a bad boy." Having, myself, met a remarkable number of bad boys, it might seem that either our experiences were drastically different or we were using the word "bad" differently. I suspect neither is true, but rather that the Father was appraising the "inner man," while I, in fact, do not acknowledge the existence of inner people.

Since we psychoanalysts have unwittingly contributed to this confusion, let one, at least, attempt a small rectifying effort. Psychoanalytic data—which should be viewed as supplementary information—is, unfortunately, often viewed as alternative (and superior) explanation. This has led to the prevalent tendency to think of the "inner" man as the real man and the outer man as an illusion or pretender.

While psychoanalysis supplies us with an incredibly useful tool for explaining the motives and purposes underlying human behavior, most of this has little bearing on the moral nature of that behavior.

Like roentgenology, psychoanalysis is a fascinating, but relatively new, means of illuminating the person. But few of us are prepared to substitute an X-ray of Grandfather's head for the portrait that hangs in the parlor. The inside of the man represents another view, not a truer one. A man may not always be what he appears to be, but what he appears to be is always a significant part of what he is. A man is the sum total of *all* his behavior. To probe for unconscious determinants of behavior and then define him in their terms exclusively, ignoring his overt behavior altogether, is a greater distortion than ignoring the unconscious completely.

Kurt Vonnegut has said, "You are what you pretend to be," which is simply another way of saying, you are what we (all of us) perceive you to be, not what you think you are.

Consider for a moment the case of the ninety-year-old man on his deathbed (surely the Talmud must deal with this?) joyous and relieved over the success of his deception. For ninety years he has affected courtesy, kindness, and generosity—suppressing all the malice he knew was within him while he calculatedly and artificially substituted grace and charity. All his life he had been fooling the world into believing he was a good man. This "evil" man will, I predict, be welcomed into the Kingdom of Heaven.

Similarly, I will not be told that the young man who earns his pocket money by mugging old ladies is "really" a good boy. Even my generous and expansive definition of goodness will not accommodate that particular form of self-advancement.

It does not count that beneath the rough exterior he has a heart—or, for that matter, an entire innards—of purest gold, locked away from human perception. You are for the most part what you seem to be, not what you would wish to be, nor, indeed, what you believe yourself to be.

Spare me, therefore, your good intentions, your inner sensitivities, your unarticulated and unexpressed love. And spare me also those tedious psychohistories which—by exposing the goodness inside the bad man, and the evil in the good—invariably establish a vulgar and perverse egalitarianism, as if the arrangement of what is outside and what inside makes no moral difference.

Saint Francis may, in his unconscious, indeed have been compensating for, and denying, destructive, unconscious Oedipal impulses identical to those which Attila projected and acted on. But the similarity of the unconscious constellations in the two men matters precious little, if it does not distinguish between them.

I do not care to learn that Hitler's heart was in the right place. A knowledge of the unconscious life of the man may be an adjunct to understanding his behavior. It is *not* a substitute for his behavior in describing him.

The inner man is a fantasy. If it helps you to identify with one, by all means, do so; preserve it, cherish it, embrace it, but do not present it to others for evelution or consideration, for excuse or exculpation, or, for that matter, for punishment or disapproval.

Like any fantasy, it serves your purpose alone. It has no standing in the real world which we share with each other. Those character traits, those attitudes, that behavior—that strange and alien stuff sticking out all over you—*that's the real you!*

## Rough draft

An Analysis of "What You See Is the Real You"

There are some things that I don't agree with in it but on the whole, Willard Gaylin's essay, "What You See Is the Real You", is a much-needed corrective to much falty thinking today. First of all, as Gaylin himself indicates to the reader, he is a psychoanalyst who is not in full agreement with the ~~curent~~ practice of his perfession. The reader sees with some surprise that he apparently believes that in his own field, as well as in other fields, there is a tendency to excuse those who act irresponsibly or even criminally, because, as the theory goes, deep down they may be decent and caring people. Gaylin informs the reader that psychoanalysis is good but studying and empha- sizing the "inner man," to use Gaylin's own term, has had the effect of relieving individuals of the moral responsibility of their actions.

I think that he wrote this article, because he wanted to tell readers that despite all this bull about criminals and other undesireables being that way because of the way they were brought up and that beneath their badness is something good just waiting to be brought out, people are really responsible for their actions. This is what this essay really says to me. You can't put the blame on environment or bad treatment or neglect or poverty or abuse or anything else. You are responsible for your actions, my friend. Period.

One of the most engenuous things about the essay is the way the author gets the reader to think this way about the need of individual responsibility. The readers gets so fascinated with the idea of the "inner man" that he/she is subtly brought around to the author's way of thinking. I have noticed this technique in several of the essays we have studied this term. An author will manipulate the reader by arguing along one line with the purpose of accomplishing something

else.   The essay by Swift that we studied is a good
example of this.⟩ But anyway, the reader can see that
Gaylin in engenuous in his use of the argument against
the reality of "inner people," because in actuality he
is not so much concerned with debating this psychological and philosophical subject as he is with finding a way to stress to the reader that it is both
inaccurate and counterproductive to blame unacceptable
behavior on one's environment.   Or on any other fac
tors for that matter.   Gaylin denies the existence of
an inner self in order to emphasize to the reader that
there is no aspect of our being more basic and fundamental, more important, than our conduct.   Gaylin
urges that we integrate our sense of self with our
behavior and think of them as one by doing so, readers
will realize that they are responsible for their own
behavior.   History, the reader tends to agree, has
certainly proven Gaylin correct in his insistence on
personal responsiblility.   The reader can just imagine
the chaos that would have resulted down through the
centuries if budding civilizations had taken a kind of
no-fault reasoning.   Failing to expect people to abide

by laws and become productive citizens, forgiving them
for all sorts of antisocial conduct, because of professed good intentions or disadvantaged upbringing.

weak

I believe that some aspects of Gaylin's argument,
however, do present problems. He goes too far. In an
effort to make his point, he exaggerates and even envolves himself in contradictions. I think the old man
is a good example of this. Gaylin's praise amounts to
praise of hypocrisy. In addition, Gaylin goes too far
in denying that a heart of gold may belie a deceptive
exterior. Readers have probably known a person who
most of the time acts selfishly but who on occasion
behaves in such a way to suggest that there is real
goodness in his nature. I believe Gaylin seems to
question that there ever could be such a person and in
doing so he fails in this part of his essay to be
convincing.

Although the essay may be logically flawed, it
appeals to the reader nevertheless as making a valit
and powerful point that should not be obscured by an
exaggeration or a contradiction here and there. We
act as we do because of what we are, and we must take

full responsibility for our acts.  It is not who or
what we think we are that counts but how we behave.

A rather serious question, however, arises to
the reader from his saying early in the essay that
he does not acknowledge the existence of inner people
but then he procedes as if he does, for he compares
the "inner man" to an X-ray, real but distinctively
different from a portrait.  I would like to ask the
question, if the inner self is a "fantasy," as he says
later, then how can it be X-rayed?

## ■ Exercise 2

*Make two lists, one that notes what is good in the above draft and one
that points out problems and errors (weaknesses that you would
work on if you were revising the paper).*

## **41g** Consider your draft unfinished.

Resist the temptation to think of your draft as the finished product.
With the final words of your draft, you may feel such relief and
satisfaction that you have no inclination to go any further. You may
even read over quickly what you have written and find it, as your
ego tells you, surprisingly good. Beware of that voice. Often the
difference between an effective writer and a poor one is the will-
ingness to revise, to change what was so hard to express, to delete
that which resulted from hard thinking, to subtract when one had
such a difficult time adding.

Your most valuable asset as a writer is a critical eye for examining your own draft. Work at standing apart from it and imagining that it belongs to someone else. Make a list of the major weaknesses in it. Do not consider it untouchable; it is but the product of a stage in the composition process, not the end result.

Follow the process of revision in section **41h** to see how the student shaped the draft on pp. 297–301 into a good paper.

# **41h**   Revise the draft.

Generally it is a good idea to take care of the larger matters (sometimes called **global revisions**) first. When you revise globally, you delete irrelevant materials, add necessary passages, and move sentences or whole sections around. Note these larger revisions in the draft below. The second paragraph has been deleted because it is badly written and repeats points suggested elsewhere. The first several lines of the third paragraph have been cut because they contain material that is not relevant to the subject. Words have been added in the fourth paragraph to make clear the reference to the "old man." The final paragraph is anticlimactic, but it will work well if moved to a new position, as indicated.

## Draft with global revisions

An Analysis of "What You See Is the Real You"

There are some things that I don't agree with in it but on the whole, Willard Gaylin's essay, "What You See Is the Real You", is a much-needed corrective to much falty thinking today. First of all, as Gaylin himself indicates to the reader, he is a psychoanalyst who is not in full agreement with the curent practice

of his perfession.   The reader sees with some surprise

that he apparently believes that in his own field as

well as in other fields, there is a tendency to excuse

those who act irresponsibly or even criminally,

because, as the theory goes, deep down they may be

decent and caring people.  Gaylin informs the reader

that psychoanalysis is good but studying and empha-

sizing the "inner man," to use Gaylin's own term, has

had the effect of relieving individuals of the moral

responsiblity of their actions.

I think that he wrote this article, because he

wanted to tell readers that despite all this bull

about criminals and other undesireables being that way

because of the way they were brought up and that

beneath their badness is something good just waiting

to be brought out, people are really responsible for

their actions.   This is what this essay really says to

me.   You can't put the blame on environment or bad

treatment or neglect or poverty or abuse or anything

else.   You are responsible for your actions, my

friend.   Period.

*delete*

One of the most engenuous things about the essay is the way the author gets the reader to think this way about the need of individual responsibility. The reader gets so fascinated with the idea of the "inner man" that he/she is subtly brought around to the author's way of thinking. I have noticed this technique in several of the essays we have studied this term. An author will manipulate the reader by arguing along one line with the purpose of accomplishing something else. The essay by Swift that we studied is a good example of this. But~~anyway,~~ the reader can see that
*delete*

¶ Gaylin is engenuous in his use of the argument against the reality of "inner people," because in actuality he is not so much concerned with debating this psychological and philosophical subject as he is with finding a way to stress to the reader that it is both inaccurate and counterproductive to blame unacceptable behavior on one's environment. Or on any other factors for that matter. Gaylin denies the existence of an inner self in order to emphasize to the reader that there is no aspect of our being more basic and fundamental, more important, than our conduct. Gaylin urges that we integrate our sense of self with our behavior and think of them as one by doing so, readers will realize that they are responsible for their own behavior. History, the reader tends to agree, has certainly proven Gaylin correct in his insistence on personal responsibility. The reader can just imagine the chaos that would have resulted down through the centuries if budding civilizations had taken a kind of no-fault reasoning. Failing to expect people to abide by laws and become productive citizens, forgiving them for all sorts of antisocial conduct, because of professed good intentions or disadvantaged upbringing.

I believe that some aspects of Gaylin's argument, however, do present problems. He goes too far. In an effort to make his point, he exaggerates and even envolves himself in contradictions. ~~I think the old man is a good example of this.~~ ~~Gaylin's praise~~ amounts to praise of hypocrisy. In addition, Gaylin goes too far in denying that a heart of gold may belie a deceptive exterior. Readers have probably known a person who most of the time acts selfishly but who on occasion

*His apparent admiration for the old man who had lived a life of deception—thinking one way and acting another—*

behaves in such a way to suggest that there is real goodness in his nature.  I believe Gaylin seems to question that there ever could be such a person and in doing so he fails in this part of his essay to be convincing.

*Put last paragraph here.* ←

Although the essay may be logically flawed, it appeals to the reader nevertheless as making a valit and powerful point that should not be obscured by an exaggeration or a contradiction here and there.  We act as we do because of what we are, and we must take full responsibility for our acts.  ~~It is not who or what we think we are that counts but how we behave.~~

*Conclusion pnt.*

A rather serious question, however, arises to the reader from his saying early in the essay that he does not acknowledge the existence of inner people but then he procedes as if he does, for he compares the "inner man" to an X-ray, real but distinctively different from a portrait.  I would like to ask the question, if the inner self is a "fantasy," as he says later, then how can it be X-rayed?

*reader*

After global revisions, you can turn to **local** (or **sentence-level**) **revisions**—that is, the correcting of errors in spelling, punctuation, and mechanics, and the clearing up of problems with wordiness, repetition, diction, and so forth. Study the corrections that have been made in the following draft. A new title has been supplied. (Avoid expressions like "A Reading of" and "An Analysis of" in titles. See **42b.**) The style has been tightened throughout the essay by deleting useless references to "the reader" (see **42f**) and expressions using the first-person pronoun *I.* Note other changes to correct repetition and redundancy, misspellings, and misuse of commas.

## Draft with local revisions

*Conduct and the Inner Self*
~~An Analysis of "What You See Is the Real You"~~

~~There are some things~~ that ~~I don't agree with in~~ ~~it but~~ Ó̸n the whole, Willard Gaylin's essay/ "What You See Is the Real You"/ is a much-needed corrective to
*faulty*
~~much~~ ~~falty~~ thinking today. ~~First of all,~~ A̸s Gaylin himself indicates ~~to the reader~~, he is a psychoanalyst
*current*
who is not in full agreement with the ~~current~~ practice
*profession*
of his ~~perfession~~. ~~The reader sees with some surprise~~
*H*
~~that~~ h̸e apparently believes that in his own field as
*s*
well as in other ~~fields~~ there is a tendency to excuse those who act irresponsibly or even criminally/ because, as the theory goes, deep down they may be

decent and caring people. ~~Gaylin informs the reader~~
~~that psychoanalysis is good but~~ $\overset{S}{\cancel{s}}$tudying and emphasizing the "inner man," to use Gaylin's ~~own~~ term, has had
the effect of relieving individuals of the moral
responsiblity of their actions.

    Gaylin is $\underset{\text{(underlined)}}{\overset{ingenious}{\underline{\text{engenuous}}}}$ in his use of the argument
against the reality of "inner people$\overset{\bullet}{,}$" ~~because~~ $\overset{I}{\cancel{i}}$n
actuality he is not so much concerned with debating
this psychological and philosophical subject as he is
with finding a way to stress ~~to the reader~~ that it is
~~both~~ inaccurate and counterproductive to blame unac-
ceptable behavior on one's environment$\cancel{,}$ $\overset{o}{\cancel{O}}$r on any
other factors $\overset{\bullet}{~}$ ~~for that matter.~~ Gaylin denies the
existence $\underset{\wedge}{}$ of an inner self in order to emphasize ~~to~~
~~the reader~~ that there is no aspect of our being more
basic ~~and fundamental~~, more important, than our con-
duct. $\overset{He}{\cancel{\text{Gaylin}}}$ urges that we integrate our sense of
self with our behavior and think of them as one $\underset{\wedge}{\overset{\bullet B}{\cancel{by}}}$
doing so, $\overset{people}{\cancel{\text{readers}}}$ will realize that they are responsible for their own behavior. ¶History$\cancel{,}$ ~~the reader tends~~
~~to agree,~~ has certainly proven Gaylin correct in his
insistence on personal responsibility. $\overset{One}{\cancel{\text{The reader}}}$ can

~~just~~ imagine the chaos that would have resulted down
through the centuries if budding civilizations had
*practiced*
~~taken~~ a kind of no-fault reasoning ~~,~~ f̶ailing to expect
people to abide by laws and become productive citizens,
forgiving them for ~~all sorts of~~ antisocial conduct,
because of professed good intentions or disadvantaged
upbringing.

S
~~I believe that~~ s̶ome aspects of Gaylin's argument,
however, do present problems. ~~He goes too far.~~ In an
*in-*
effort to make his point, he exaggerates and even ~~en-~~
*volves*
~~volves~~ himself in contradictions. His apparent admiration for the old man who had lived a life of decep-
tion--thinking one way and acting another--amounts to
praise of hypocrisy. In addition, Gaylin goes too far
*underlie*
in denying that a heart of gold may ~~belie~~ a deceptive
*Almost everyone has*
exterior. ~~Readers have probably~~ known a person who
most of the time acts selfishly but who on occasion
behaves in such a way to suggest that there is real
*or her*
goodness in his ∧ nature. ~~I believe~~ Gaylin seems to
question that there ever could be such a person, and in
doing so he fails in this part of his essay to be
convincing.

*somewhat more*
A ~~rather~~ serious question, ~~however,~~ arises ~~to the reader~~ from his saying early in the essay that he does not "acknowledge the existence of inner people" but then *proceeding* ~~he procedes~~ as if he does, for he compares the "inner man" to an X-ray, real but distinctively different from a portrait. ~~I would like to ask the question,~~ If the inner self is a "fantasy," as he says later, then how can it be X-rayed?

Although the essay may be logically flawed, it *makes* ~~appeals to the reader nevertheless as making~~ a *valid* ~~valit~~ and powerful point that should not be obscured by an exaggeration or a contradiction here and there. We act as we do because of what we are, and we must take full responsibility for our acts. It is not who or what we think we are that counts but how we behave.

Before submitting your paper, read it over two or three times, at least once aloud. Listen for annoying repetitions. Compare the finished copy of the paper "Conduct and the Inner Self" on pp. 312–315 to the draft on pp. 297–301.

You will find the following checklist helpful during the process of writing and revising as well as just before turning in your paper.

## Checklist

### Title

The title should accurately suggest the contents of the paper.

It should attract interest without being excessively novel or clever.

It should not be too long.

NOTE: Do not underline the title of your own paper, and do not put quotation marks around it.

### Introduction

The introduction should be independent of the title; no pronoun or noun in the opening sentence should depend for meaning on the title.

It should create interest.

It should properly establish the tone of the paper as serious, humorous, ironic, or otherwise.

It should include a thesis statement that declares the subject and the purpose directly but at the same time avoids worn patterns like "It is the purpose of this paper to. . . ."

### Body

The materials should develop the thesis statement.

The materials should be arranged in logical sequence.

Strong topic sentences (see **40a**) should clearly indicate the direction in which the paper is moving and the relevance of the paragraphs to the thesis statement.

Technical terms should be explained.

Paragraphs should not be choppy.

Adequate space should be devoted to main ideas; minor ideas should be subordinated.

Concrete details should be used appropriately; insignificant details should be omitted.

### Transitions

The connections between sentences and those between paragraphs should be shown by appropriate linking words and by repetition of parallel phrases and structures (see **40f**).

### Conclusion

The conclusion should usually contain a final statement of the underlying idea, an overview of what the paper has demonstrated.

The conclusion may require a separate paragraph; but if the paper has reached significant conclusions all along, such a paragraph is not necessary for its own sake.

The conclusion should not merely restate the introduction.

### Proofreading

Allow some time, if possible at least one day, between the last draft of the paper and the final, finished copy. Then you can examine the paper objectively for wordiness, repetition, incorrect diction, misspellings, faulty punctuation, choppy sentences, vague sentences, lack of transitions, and careless errors.

## **41i** Model paper

Conduct and the Inner Self

On the whole, Willard Gaylin's essay "What You See Is the Real You" is a much-needed corrective to faulty thinking today. As Gaylin himself indicates, he is a psychoanalyst who is not in full agreement with the current practice of his profession. He apparently believes that in his own field as well as in others,

there is a tendency to excuse those who act irresponsibly or even criminally because, as the theory goes,
deep down they may be decent and caring people.
Studying and emphasizing the "inner man," to use
Gaylin's term, has had the effect of relieving individuals of the moral responsiblity of their actions.

Gaylin is ingenious in his use of the argument
against the reality of "inner people." In actuality,
he is not so much concerned with debating this psychological and philosophical subject as he is with finding a way to stress that it is inaccurate and counter
productive to blame unacceptable behavior on one's
environment or any other factors.  He denies the existence of an inner self in order to emphasize that
there is no aspect of our being more basic or important than our conduct.  He urges that we integrate our
sense of self with our behavior and think of them as
one. By doing so, people will realize that they are
responsible for their own behavior.

History has certainly proven Gaylin correct in his
insistence on personal responsibility.  One can imagine
the chaos that would have resulted down through the

centuries if budding civilizations had practiced a kind of no-fault reasoning, failing to expect people to abide by laws and become productive citizens, forgiving them for antisocial conduct because of professed good intentions or disadvantaged upbringing.

Some aspects of Gaylin's argument, however, do present problems. In an effort to make his point, he exaggerates and even involves himself in contradictions. His apparent admiration for the old man who had lived a life of deception--thinking one way and acting another--amounts to praise of hypocrisy. In addition, Gaylin goes too far in denying that a heart of gold may underlie a deceptive exterior. Almost everyone has known a person who most of the time acts selfishly but who on occasion behaves in such a way to suggest that there is real goodness in his or her nature. Gaylin seems to question that there ever could be such a person, and in doing so he fails in this part of his essay to be convincing.

A somewhat more serious question arises from his saying early in the essay that he does "not acknowledge the existence of inner people" but then proceeding as if he does, for he compares the "inner man" to an X-ray, real but distinctively different from a portrait. If the inner self is a "fantasy," as he says later, then how can it be X-rayed?

Although the essay may be logically flawed, it makes a valid and powerful point that should not be obscured by an exaggeration or contradiction here and there. We act as we do because of what we are, and we must take full responsibility for our acts. It is not who or what we think we are that counts but how we behave.

### Exercise 3

■ *Write a critique (500–600 words) of either Willard Gaylin's essay "What You See Is the Real You" (pp. 295–297) or the model paper "Conduct and the Inner Self" (pp. 312–315). Evaluate the strengths, and point out any weaknesses in the author's argument.*

## 41j  Writing about yourself

Henry David Thoreau wrote in *Walden* that he would not have written so much about himself if he had known another subject as well. Your instructor may ask you to write a paper about yourself or about a personal experience—something that you consider lasting and important in your life—in other words, about a subject that no one knows as much about as you. Unlike the model paper on pp. 312–315, in which the writer becomes more effective as he or she avoids personal intrusions (see **42g**), this kind of essay by necessity draws you into the spotlight. It provides you with the opportunity to sort out your experience, to remember creatively, to connect early images and details with later meanings, and to write something that will be not only engaging and refreshing to others but significant to you. The act of writing such papers may teach you something about yourself.

Writing about yourself can be the easiest or the most difficult task you undertake in a composition course. It can be easy if you merely relate something that happened to you and tell a little about your emotions at the time or if you simply describe superficially a relative or friend whom you admire, filling your paper with commonplaces and clichés. Avoid the temptation to write about how you have been inspired to fulfill your ambitions or about your dream to become this or that. Much more difficult is probing your heart and consciousness to assess what a person or an experience has meant to you. It takes time. It requires effort, thought, and revisions; and sometimes it is painful.

The *challenging* paper about yourself is the only kind worth writing. It is challenging to make yourself interesting to others, to reveal enough to show that you are in the mainstream of humanity but not enough to come across as either self-pitying or arrogant, to deal with a simple event—something thousands of others have experienced—in such a way, with genuineness and restrained emotion, that you make others feel something of what you feel. It may take you a while to become good at writing about yourself. If you do become good at it, you may become good at many other things, too.

Below is a paper written in response to an assignment in which students were asked to write about themselves.

What the Ocean Means to Me

I do not know how old a child has to be in order for events to start making the kind of impression that constitutes memory. Most people can remember from age five or six, and that is about how old I was when I first saw an ocean, the Atlantic Ocean, and it is the very first thing that I can remember about my life-- that image of the ocean. Everything else that I can remember about my childhood comes after that. Scientists say that life on this planet came from the ocean. In a way the scientists do not mean, my life comes out of the ocean.

My family lives inland amid the foothills of a large mountain range. There is little enthusiasm among my parents and brothers and sisters for living anywhere else. My mother has told me that it was almost by accident that we went to Florida that winter; it certainly was no vacation. My father was looking into an employment opportunity and decided to take the family with him just for the weekend. Before checking into a motel, my father took us straight to the beach for a look.

That is where my memory begins.  It was not a sunny, bright day with people romping on the beach and with colorful umbrellas and beach balls.  It was cold and windy.  The beach was deserted.  If there had been people near the ocean, maybe the scene would not have made such an impression on me, but we were alone.  The waves were high with the wind; it was a rough sea.  I suppose I was the only one in the family--being the youngest--who had not previously seen the ocean.  The rest of them appeared anxious to watch for my reaction.  Externally, it may not have been much, but internally, it cut a deep groove in my memory.

As I think about it now, what I felt was a sequence of emotions.  Bewilderment and fear came over me as I looked out as far as I could see and as far as possible to the right and left.  I remember wanting to cry and run away.  My father held my hand and watched me, so I stuck with it.  No one else seemed to be afraid, so I took heart from them.  But it was the ocean itself, rather than my family, that calmed me.  Somehow I seemed to sense after the first few moments that its vastness was no threat to me but a comfort. I will not say that it spoke to me because it did not, but in the calmer state I began to feel something that I remember vividly: that this ocean belonged there, that it was right for it to be there. So everything suddenly seemed all right to me. When you see something that belongs where it is and it is right for it to be there, fear goes away.

Since I first saw the ocean, I have returned to it at every opportunity. I nearly drowned in it when I was about twelve. My brother pulled me from a strong current that was carrying me out. I was with a friend of my high-school years when he was bitten on the calf by a shark. We were only about waist-deep in the water, which suddenly turned red with his blood as he hurried--white-faced--to reach the shore. His leg required seventy stitches.

I am not sentimental about the sea. It is filled with danger. But I am drawn to it with the same emotions that I experienced upon first seeing it. I knew then and I know now that it was not just placed there recently. It belongs there, and it has been there a

```
long time.   I have changed a lot since I first saw
it, but it has not changed at all.   Everyone needs in
the memory an image of something that is permanent.
For me, it is the ocean.   Whenever I think about it,
I feel that everything is all right.
```

■ **Exercise 4**

*Some of the pitfalls of personal writing are shallowness, dullness, sentimentality, and self-centeredness. Write a page in which you show how the author of "What the Ocean Means to Me" avoids such pitfalls or falls into them.*

■ **Exercise 5**

*Compose an essay of approximately 600 words about an experience that has significantly influenced your thinking.*

## **41k**  Learn from your mistakes.

The process of composition is not complete until you have carefully examined your paper after it is returned to you, looked up your errors in this book and in a dictionary so that you understand how to correct them, and revised along the lines suggested by your instructor. Learn from your mistakes so that the same problems do not turn up in future papers.

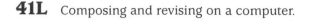

# **41L**   Composing and revising on a computer.

Most of the writing techniques discussed in this chapter—for example, listing and freewriting—can be effectively practiced on a personal computer with a word-processing program. Many writers find it easier to initiate the flow of ideas on a computer than on a typewriter or with pen and paper. Thoughts can be recorded on the screen and discarded or retained quickly and easily. The popularity of computers is based not merely on their ability to save time; they enable users to engage in the process of creating without seriously interrupting the flow of mental current.

If you are used to writing drafts of papers in longhand or on a typewriter, you probably have a built-in resistance to throwing away sheets of paper and beginning all over again, recopying, retyping, and laboriously correcting mistakes. The awareness of inconvenience often acts as a psychological barrier to thorough revisions of drafts. With a computer you can insert words, sentences, and paragraphs, make minor or extensive changes in wording, delete passages, check spelling, find good synonyms, and revise in other ways with an ease undreamed of several years ago. Use these ways of making changes, and cultivate the habit of thorough revision, the best habit a writer can develop.

Though by no means complete, the following list will furnish some helpful suggestions about writing on a computer.

### *Save.*

Use the **Save** command to protect yourself against losing what you have written. Some mistake—hitting the wrong keys, disconnecting or cutting off the machine—a power surge, or other mishap could erase hours of work. Different versions of a paper can be saved under different titles and then profitably compared during the process of revision.

### Print out.

Regularly print out copies of what you are working on. Seeing on paper what you have been viewing on the screen gives you another perspective and makes it easier to spot errors. In addition, if you print out several versions of a document, you may find in an early draft valuable material that you can use, material that would be lost if you simply made all changes on the computer without periodically printing out.

### Back up.

Computers do break down; therefore, it is good insurance against losing everything you have written to use a back-up program if you have one, or simply to copy material that you have on the hard disk of your computer onto a diskette (floppy disk) to be stored separately.

### Explore.

Most current word-processing programs include a fascinating variety of functions. Learn as much as you can as early as you can about the software you are using. Being thoroughly familiar with the particular computer you are using and with its word-processing program will save you time later on and increase your effectiveness and pleasure.

### Work comfortably.

Prepare a comfortable working place with easy access to all the materials you will need.

### Take frequent breaks.

Sitting at a computer for hours without rest can cause problems with your eyes and neck and make you a less efficient writer.

### Replace ribbons regularly.

It is easy to put off changing the ribbon or cartridge on your printer, but turning in a paper with faint type is inconsiderate and sloppy.

### Store disks carefully.

Preserve what you have placed on floppy disks by keeping them where they will not be exposed to heat, dampness, or a magnetic field.

# Literature

42 ■ Writing About Literature, 326

# 42 Writing About Literature

Writing about novels, short stories, poems, and plays differs from the composition of other types of papers. The way you approach your subject should reflect your awareness of this difference. The model paper on pp. 312–315 deals almost exclusively with what Willard Gaylin says and with the student writer's agreements and disagreements with him. Ideas occupy center stage—as they should in this kind of writing. A poem or a short story, however, is more than a collection of ideas intelligently and clearly expressed; it is the presentation of ideas and emotions in a medium of art involving a high degree of creative imagination and technical skill.

If you write about a novel or a play in precisely the same way as you approach works dealing with history or politics or sociology, stressing only what the author has "to say," you are ignoring what makes artistic literature different. Concentrate not only on what the author has to say but also on *how* he or she says it. Show how the ideas grow out of the art. The craft of poetry, fiction, and drama is at least as important as the ideas they project. If your audience cannot tell from your approach that you are writing about art (rather than about economics, philosophy, or popular culture), if—from the way you approach your subject—the book or essay that you are writing about could just as well be a treatise on Puritan values as Hawthorne's novel *The Scarlet Letter,* you need to redesign your paper so that you devote more attention to literature as literature.

Express your reactions to literature in a way that does justice to the richness and complexity of the work. Extensive subjectivity is a danger. You should state views, but remember that you have an obligation not only to express but also to convince. Look for solid evidence from the work to corroborate your opinions. Works of literature can support multiple interpretations, but some readings are more convincing than others, and some can be shown to be altogether erroneous.

**42a**  Choose a literary work that interests you. Write about the feature of the work that interests you most.

Your instructor may assign you a particular work for a paper, or you may be allowed to choose for yourself. If you have any freedom of choice, select carefully. An unwise selection will make success difficult even if you write well.

Give thought to your choice. You are likely to write a better paper if you choose a **work new to you,** one that you have never written about before. Search for the kind of topic you like. Look at familiar authors, famous authors whose works you have not read, works recommended to you by people who have tastes somewhat like yours. Look at shelves of books and lists of authors you have heard about but not read, new stories and poems by authors familiar to you, anthologies, collections. You are unlikely to write a good paper about a work that you regard as dull or mediocre. If a literary work contains no mystery for you at first, you are unlikely to explain it originally or interestingly to others.

Avoid shallow and obvious literature. If you read something that makes you think, "That is exactly what I have always known," you should doubt the originality of the author. If you think, "True, but I never thought of anything like that before," you may have a good selection on your hands. A work that interprets or explains itself will not leave much for you to write about. Good papers are often derived from literature that is somewhat puzzling and obscure during and immediately after the first reading. Additional readings and careful thought can produce surprises that give you something to write about.

Do not decide to write on a particular subject before you read the work. What you think will be a good subject may not be there at all. Find your subjects as you read and as you think. Let good subjects grow and develop; bad ones will disappear or fade away. The most surprising and significant topics may come to you when you least expect them—as you slowly wake in the morning, for example, or as you engage in some kind of physical activity.

The basic methods for planning and writing a paper about literature are the same as those you would use for writing any other kind of paper (see p. 284). List topics and subtopics; group them; arrange them; rearrange them; write whatever comes into your mind, whether it seems good or bad at the moment; write to keep the words going, even when what you are writing seems unimportant. Cross out. Put checkmarks by important points; photocopy parts of works; make notations of lines, passages, images, sentences. Write quickly for a time. Think and write slowly and carefully about crucial subjects. Stop writing and walk about. Use every physical and mental device you can think of to tease good thoughts out of your mind.

Be willing to give up ideas that are inferior. Expect surprises and twists and turns during the process of writing. Few papers are written like bullets speeding toward a single idea with a calculated aim; most of the time they act more like floods, gathering valuables and debris at the same time. After the first draft, you can begin separating bad from good ideas.

## Kinds of literary papers

Writings about literature fall into several categories. A few of the more significant ones are explained below.

### *Interpretation*

Many papers are interpretations derived mainly from close study of the literary work. An interpretative paper identifies methods and ideas. Through analysis, the writer presents specific evidence to support the argument.

Distinguish carefully between the thinking of a character and that of the author. Unless an author speaks in his or her own voice, you can deduce what the author thinks only from the work as a whole, from effects of events in the plot, and from attitudes that develop from good or bad characters. Some works of literature depict a character whose whole way of life is opposite to the author's views. To confuse the character with the author is to make a crucial mistake. Sometimes it is clear immediately what an author thinks about the characters, but not always.

### Review

A good review of a book or an article

1. identifies the author, the title, and the subject.
2. summarizes accurately the information presented and the author's argument.
3. describes and perhaps categorizes the author's methods.
4. provides support (with evidence and argument) for the good points and explains the author's mistakes, errors, misjudgments, or misinterpretations.
5. evaluates the work's accomplishments and failures.

### Character analysis

A character sketch is a tempting kind of paper to write, but a good analysis of a character or an explanation of an author's methods of characterization is truly a difficult task. This kind of paper easily turns into a superficial summary when it fails to consider motivations, development, change, and interrelationships of characters. Meaningful interpretation of a character may define something that the character does not know about himself or herself, traits that the author reveals only by hints and implications. For example, if you can show that a character clearly lacks self-understanding, you may be on the way to writing a good paper.

Of course, an analysis may *not* deal at all with *what the character is* but consider instead the *methods* the author uses to reveal the character. Body language and modes of speech, facial expressions or changes in mood, excessive talking or silence, and appearance are some of these methods. Remember that the author characterizes; the critic discusses the methods of characterization.

## Analysis of setting

Often the time and place in which a work is set reveal important moods and meanings. Setting may help to indicate the manners and emotions of characters by showing how they interact with their environments. When you write about setting, you accomplish very little by merely describing it. Show what its functions are in the literary work.

## Technical analysis

The analysis of technical elements in literature—imagery, symbolism, point of view, structure, prosody, and so on—requires special study of the technical term or concept as well as of the literary work itself. Begin by looking up the term in a good basic reference book, such as *A Handbook to Literature*. However, you should never be concerned with vocabulary and technical terms for their own sake. If you discover an aspect of a work that you wish to discuss, find the exact term for it by discussing it with your teacher, looking it up in a dictionary, and studying it in a handbook or an encyclopedia. Then determine what the technique does in the work you are writing about.

## Combined approaches

Many papers combine different approaches. A thoughtful paper on imagery, for example, does more than merely point out the images, or even the kinds of images, in the literary work; it uses the imagery to interpret, analyze, or clarify something else as well—theme, structure, characterization, mood, relationship, recurrent patterns, and so on. Depending on the subject and the work you are writing about, many aspects can mingle to accomplish a single objective.

## **42b** Give the paper a precise title.

Do not search for a fancy title at the expense of meaning. Authors of literary works often use figurative titles like *Death in the Afternoon* and *The Grapes of Wrath,* but you would be wise to designate your subject more literally. Be precise; state your subject in the title.

Do not use the title of the work as your own title, as in "Robert Frost's 'Directive.'" Do not merely announce that you are writing about a work by calling your paper "An Analysis of Frost's 'Directive'" or "An Interpretation of . . ." or "A Criticism of. . . ." The fact that you are writing the paper indicates that you are writing an analysis, a criticism, or something of the sort.

Stick to the topic named in the title. The topic sentence of every paragraph should point back to the title, the introduction, and the thesis statement.

## **42c** Do not begin with broad philosophical statements.

You do not have to create for your paper an introduction composed of sage observations about history or human nature, as a prelude to your focus upon more specific matters. In fact, such opening statements and the comments that follow them usually tend toward the trite and commonplace (see **38e**); therefore, get immediately to your subject without philosophical fanfare, which often comes across as mere padding. Avoid openings like the following:

"In today's complex world . . ."
"Human beings have always . . ."
"From the beginning of time . . ."
"In the annals of . . ."
"Throughout the past . . ."
"From the dawn of civilization . . ."

**42d** Organize and develop the paper according to significant ideas.

Do not automatically organize your paper by following the sequence of the literary work. Sometimes the result of this order can be poor topic sentences, summary rather than analysis, mechanical organization, and repetitive transitional phrases.

Usually it is better to break up your overall argument into several aspects and to move from one of these to the next.

First sentences that provide mechanical and dull information do not encourage further reading. Avoid generality, as in the following sentence:

> T. S. Eliot, in his famous poem written in 1922, *The Waste Land,* expressed a theme that has been a frequent subject of works of literature.

Instead, you might try something like this:

> The relationship between the physical, mental, and spiritual health of a ruler of a nation and the condition of the people is a subject of T. S. Eliot's poem *The Waste Land.*

The first sentence states nothing of true significance; the second announces a particular topic to be explored.

Just as a paper should not begin mechanically, the parts should not contain mechanical first sentences and dull transitions. Your paragraphs will not create interest if they use beginnings like these:

"In the first stanza . . ."

"In the following part . . ."

"At the conclusion . . ."

**42f**

## **42e**  Avoid extensive paraphrasing and plot summary.

Keep summarizing to a minimum, just enough to furnish context and to be clear. Mere plot-summary of fiction or paraphrase of poetry is inadequate. Summary must prove your argument; it should not be an end in itself. Extensive plot-summary and paraphrase are often the sure marks of writers who have not thought enough about a work of literature to do more than retell the story or slide along the surface of the poem.

NOTE:  When you do summarize, use the historical present tense (see **4a**). Tell what *happens*, not what *happened*.

## **42f**  Think for yourself.

Thinking—hard, concentrated thinking—is probably the most important single preparation for writing an effective essay. Start early enough on your assignment so that you can devote time to this essential step in the process.

You do not have to come up with an idea that no one else has ever thought of as long as you do come up with an idea, one that is original in the sense that you thought of it yourself. It may not be profound or unusual, but it is *yours*, and that is the important factor. You are developing the eye of a critic when you can read a work of literature and (after a period of brainstorming and germination) can think up your own angle (even if your instructor has assigned you a specific topic).

**42g** Write about the literature, not about yourself or "the reader."

The process by which you discovered what you are writing about is usually a dull subject to other people. Give the results of your explorations, not details of the various steps.

Your teacher will know that what you write is your opinion unless you indicate otherwise; therefore, you should not repeat expressions like "In my opinion" or "I believe that." Extensive use of first-person pronouns *(I, my)* takes the emphasis from what you are writing about.

In an effort to avoid the pronoun *I,* do not develop the annoying habit of repeating the phrase "the reader." Writing about "the reader" is logically erroneous because the expression implies readers in general, and you cannot speak for this vast, varied, and complex group. Once you begin using "the reader," it is difficult to stop, and your style will be greatly weakened by the resulting wordiness. Though you may not realize it, when you make extensive use of first-person pronouns and "the reader," you are creating an impression of self-consciousness that can annoy or even alienate.

The rough draft entitled "An Analysis of 'What You See Is the Real You'" (pp. 297–301) vividly illustrates the problems discussed above. Review that paper, noting how the student's use of "I" and "the reader" severely damages style. Then note how the faults were corrected in the model paper on pp. 312–315.

**42h** Provide evidence.

First make a specific point; then back it up with evidence: that is, selective recounting of details related to your argument, paraphrases of sections of the work, and quotations. Generalizations without evidence are seldom convincing, and your goal as a writer should be to convince.

# **42i**  Quote sparingly and effectively.

Appropriate quotations from the work of literature you are writing about are indispensable. Many times they make the best evidence, even though they may be brief. You can weave them into your argument in a number of ways, one of the most effective of which is to work them into a sentence of your own. As long as the quotation is brief enough not to cause awkwardness (see p. 111), you may place it at the beginning of your sentence, in the middle, or at the end. (For the use of commas and colons with quotations, see pp. 135, 151).

BRIEF QUOTATION AT THE BEGINNING OF A SENTENCE

"A devilish thing" is what Captain Loft calls chocolate in Steinbeck's *The Moon Is Down* (158).

BRIEF QUOTATION IN THE MIDDLE OF A SENTENCE

Captain Loft's reference to chocolate as "a devilish thing" in Steinbeck's *The Moon Is Down* comes near the end of the novel (158).

BRIEF QUOTATION AT THE END OF A SENTENCE

In Steinbeck's *The Moon Is Down*, Captain Loft calls chocolate "a devilish thing" (158).

NOTE:  In the above sentences, observe the positioning of the parentheses containing the page number. Note also punctuation in relation to quotation marks and parentheses.

Prefer brief but telling quotations over long ones. Long quotations can become boring, and they often leave the impression of haste in composition. Paraphrase is sometimes more effective than long quotations. Quote enough to make your case convincing; do not quote too often or too lengthily. If you find it necessary to quote a prose passage of over four lines, or over three lines of poetry, set the quotation off without quotation marks (see pp. 160–161).

# 42j Do not moralize.

Good criticism does not preach. Do not use your paper as a platform from which to state your views on the rights and wrongs and conditions of the world. A literary paper can be spoiled by an attempt to teach a moral lesson.

Begin your paper by introducing the work and your ideas about it as literature. If the author discusses morals, religion, or social causes, his or her literary treatment of these subjects is a proper topic. Do not begin your paper with your own ideas about the world and then attempt to fit the literature to them or merge your ideas with the author's. Think for yourself, but *think about the author's writing*.

# 42k Acknowledge your sources.

Define the difference between what other critics have written and what you think. State your contribution. Do not begin papers or paragraphs with the names of critics and their views before you have presented your own ideas.

Develop your own thesis. Stress your views, not those of others. Use sources to show that other critics have interpreted the literature correctly, to correct errors in criticism that is otherwise excellent, to show that a critic is right but that something needs to be added, or to show that no one has previously written about your subject.

It is not a problem if no critic has written about the work or the point you are making unless your instructor requires you to have a number of sources. However, it is a serious error to state incorrectly that nothing has been written about your subject. Be thorough in your investigation.

For bibliographies of writings about literature, see **43b**. For information about plagiarism and documentation, see **43f** and **43g**.

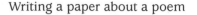

## Writing a paper about a poem

If you are allowed to select a poem to write about, choose one that appeals to your heart as well as to your head. Then, to yourself, analyze why you like it in both ways. Read it over and over, looking up words that you are uncertain about and trying to notice more with each reading. Through this process, find what you consider the center of the poem, that image or idea around which the elements of the work revolve. Then make that the central subject of your paper.

Read carefully the following poem and the model paper that follows it.

## In Mind

There's in my mind a woman
of innocence, unadorned but

fair-featured, and smelling of
apples or grass. She wears

a utopian smock or shift, her hair
is light brown and smooth, and she

is kind and very clean without
ostentation—
         but she has
no imagination.
          And there's a
turbulent moon-ridden girl

or old woman, or both,
dressed in opals and rags, feathers

and torn taffeta,
who knows strange songs—

but she is not kind.

DENISE LEVERTOV

## Model paper

Innocence and Imagination: Contrasts in
Denise Levertov's "In Mind"

Two very different women appear in the poem "In Mind" by Denise Levertov. Both are attractive in some ways, but both also have shortcomings. As the speaker of the poem describes the two women who are "in my mind" (1), she may be talking about two actual people she knows. It is more likely, however, that she is describing two halves of herself: two kinds of people that she could be or has been at different stages in her life. Through these two women, Levertov examines the contrasts between innocence and imagination.

The first part of herself the speaker considers is the "woman / of innocence" (1-2), who is characterized in simple and natural terms. She is described as "unadorned" and lacking "ostentation" (2, 8). Practicality governs this woman, who wears "a utopian smock or shift" (5) rather than any elaborate costume. Levertov notes that this woman is "fair-featured," but the only specific visual detail of her appearance given is that her hair "is light brown and smooth"

(3, 6). This description suggests a simple attrac-
tiveness but little glamour. The woman's innocence
may call to mind the Biblical Eve, a resemblance that
is further developed by the fact that she smells of
apples or grass rather than of an artificial perfume.

The generally positive description of the woman of
innocence comes to an abrupt halt, however, with a
dash and the transitional word but. Levertov concludes
the first half of the poem by noting that this woman

> is kind and very clean without
>
> ostentation--
>
>             but she has
>
> no imagination. (7-10)

Such simplicity and practicality do not encourage the
unpredictable imagination.

The second woman in the poem is neither simple nor
natural. She is a "turbulent moon-ridden girl / or
old woman, or both" (12-13). Levertov's initial
account is also turbulent or even contradictory: she
cannot even decide this woman's age. The unusual
description of her as "ridden" by the moon implies
that she is haunted by or obsessed with it--perhaps by

its beauty or association with romantic love. Her
physical appearance is also unusual. Instead of a
neat, clean smock, the second woman wears an exotic
but ragged costume: she is "dressed in opals and rags,
feathers / and torn taffeta" (14-15).

Her uncertain age, association with the moon, and
bizarre clothing contrast sharply with the character
of the woman of innocence. Levertov never uses the word
imagination to refer to the second woman, but these
descriptions demonstrate that she has a great deal of
imagination, which is further revealed by the fact
that she "knows strange songs--" (16). These songs
could refer to the creative ability to compose poems.
Yet the dash again signals that Levertov is moving to
this woman's shortcomings. The poem concludes with
the isolated line "but she is not kind" (17).

With the portraits of these two women, Levertov
explores the paradoxical contrasts between innocence
and imagination. The imagination is exotic, beautiful, and unpredictable, but it is also turbulent,
strange, and rude. The woman of imagination, however,
is the one who writes poems. The practical woman

lacks imagination, but she is clean and fair, with her own simple beauty. She may also practice kindness and consideration for others.

The speaker of the poem seems to have found both of these women within herself at different times. Their differing ages suggest that imaginative stages of life--during youth or old age--may be complemented by the innocence and kindness of middle age. Levertov thus affirms both innocence and imagination in her brief but thoughtful poem "In Mind."

## Writing a paper about a work of fiction

Read the following short story carefully.

### On the U.S.S. *Fortitude*

Ron Carlson

Some nights it gets lonely here on the U.S.S. *Fortitude.* I wipe everything down and sweep the passageways, I polish all the brass and check the turbines, and I stand up here on the bridge charting the course and watching the stars appear. This is a big ship for a single-parent family, and it's certainly better than our one small room in the Hotel Atlantis, on West Twenty-second Street. There the door wouldn't close and the window wouldn't open. Here the kids have room to move around, fresh sea air, and their own F/A-18 Hornets.

I can see Dennis now on the radar screen. He's out two hundred miles and closing, and it looks like he's with a couple of friends. I'll be able to identify them in a moment. I worry when Cherry doesn't come right home when it starts to get dark. She's only twelve. She's still out tonight, and here it is almost twenty-one hundred hours. If she's gotten vertigo or had to eject into the South China Sea, I'll just be sick. Even though it's summer, that water is cold.

There's Dennis. I can see his wing lights blinking in the distance. There are two planes with him, and I'll wait for his flyby. No sign of Cherry. I check the radar: nothing. Dennis's two friends are modified MIGs, ugly little planes that roar by like the A train, but the boys in them smile and I wave thumbs up.

These kids, they don't have any respect for the equipment. They land so hard and in such a hurry—one, two, three. Before I can get below, they've climbed out of their jets, throwing their helmets on the deck, and are going down to Dennis's quarters. "Hold it right there!" I call. It's the same old story. "Pick up your gear, boys." Dennis brings his friends over—two nice Chinese boys, who smile and bow. "Now, I'm glad you're here," I tell them. "But we do things a certain way on the U.S.S. *Fortitude.* I don't know what they do where you come from, but we pick up our helmets and we don't leave our aircraft scattered like that on the end of the flight deck."

"Oh, Mom," Dennis groans.

"Don't 'Oh, Mom' me," I tell him. "Cherry isn't home yet, and she needs plenty of room to land. Before you go to your quarters, park these jets below. When Cherry gets here, we'll have some chow. I've got a roast on."

I watch them drag their feet over to their planes, hop in, and begin to move them over to the elevator. It's not as if I asked him to clean the engine room. He can take care of his own aircraft. As a mother, I've learned that doing the right thing sometimes means getting cursed by your kids. It's O.K. by me. They can love me later. Dennis is not a bad kid; he'd just rather fly than clean up.

Cherry still isn't on the screen. I'll give her fifteen minutes and then get on the horn. I can't remember who else is out here. Two weeks ago, there was a family from Newark on the U.S.S. *Tenth*

*Amendment,* but they were headed for Perth. We talked for hours on the radio, and the skipper, a nice woman, told me how to get stubborn skid marks off the flight deck. If you're not watching, they can build up in a hurry and make a tarry mess.

I still hope to run across Beth, my neighbor from the Hotel Atlantis. She was one of the first to get a carrier, the U.S.S. *Domestic Tranquillity,* and she's somewhere in the Indian Ocean. Her four girls would just be learning to fly now. That's such a special time. We'd have so much to talk about. I could tell her to make sure the girls always aim for the third arresting wire, so they won't hit low or overshoot into the drink. I'd tell her about how mad Dennis was the first time I hoisted him back up, dripping like a puppy, after he'd come in high and skidded off the bow. Beth and I could laugh about that—about Dennis scowling at his dear mother as I picked him up. He was wet and humiliated, but he knew I'd be there. A mother's job is to be in the rescue chopper and still get the frown.

I frowned at my mother plenty. There wasn't much time for anything else. She and Dad had a little store and I ran orders and errands, and I mean ran—time was important. I remember cutting through the Park, some little bag of medicine in my hand, and watching people at play. What a thing. I'd be taking two bottles of Pepto-Bismol up to Ninety-first Street, cutting through the Park, and there would be people playing tennis. I didn't have time to stop and figure it out. My mother would be waiting back at the store with a bag of crackers and cough medicine for me to run over to Murray Hill. But I looked. Tennis. Four people in short pants standing inside that fence, playing a game. Later, I read about tennis in the paper. But tennis is a hard game to read about at first, and it seemed a code, like so many things in my life back then, and what did it matter, anyway? I was dreaming, as my mother was happy to let me know.

But I made myself a little promise then, and I thought about it as the years passed. There was something about tennis—playing inside that fence, between those lines. I think at first I liked the idea of limits. Later, when Dennis was six or so and he started going down the block by himself, I'd watch from in front of the

Atlantis, a hotel without a stoop—without an entryway or a lobby, really—and I could see him weave in and out of the sidewalk traffic for a while, and then he'd be out of sight amid the parked cars and the shopping carts and the cardboard tables of jewelry for sale. Cherry would be pulling at my hand. I had to let him go, explore on his own. But the tension in my neck wouldn't release until I'd see his red suspenders coming back. His expression then would be that of a pro, a tour guide—someone who had been around this block before.

If a person could see and understand the way one thing leads to another in this life, a person could make some plans. As it was, I'd hardly even seen the stars before, and now here, in the ocean, they lie above us in sheets. I know the names of thirty constellations, and so do my children. Sometimes I think of my life in the city, and it seems like someone else's history, someone I kind of knew but didn't understand. But these are the days; a woman gets a carrier and two kids in their Hornets and the ocean night and day, and she's got her hands full. It's a life.

And now, since we've been out here, I've been playing a little tennis with the kids. Why not? We striped a beautiful court onto the deck, and we've set up stanchions and a net. I picked up some racquets three months ago in Madagascar, vintage T-2000s, which is what Jimmy Connors used. When the wind is calm we go out there and practice, and Cherry is getting quite good. I've developed a fair backhand, and I can keep the ball in play. Dennis hits it too hard, but what can you do—he's a growing boy. At some point, we'll come across Beth, on the *Tranquillity,* and maybe all of us will play tennis. With her four girls, we could have a tournament. Or maybe we'll hop over to her carrier and just visit. The kids don't know it yet, but I'm learning to fly high-performance aircraft. Sometimes when they're gone in the afternoons, I set the *Fortitude* into a wind at thirty knots and practice touch and go's. There is going to be something on Dennis's face when he sees his mother take off in a Hornet.

Cherry suddenly appears at the edge of the radar screen. A mother always wants her children somewhere on that screen. The radio crackles, "Mom. Mom. Come in, Mom." Your daughter's voice, always a sweet thing to hear. But I'm not going to pick up right away. She can't fly around all night and get her old mom just like that.

"Mom, on the *Fortitude*. Come in, Mom. This is Cherry. Over."

"Cherry, this is your mother. Over."

"Ah, don't be mad." She's out there seventy-five, a hundred miles, and she can tell I'm mad.

"Cherry, this is your mother on the *Fortitude*. You're grounded. Over."

"Ah, Mom! Come on. I can explain."

"Cherry, I know you couldn't see it getting dark from ten thousand feet, but I also know you're wearing your Swatch. You just get your tail over here right now. Don't bother flying by. Just come on in and stow your plane. The roast has been done an hour. I'm going below now to steam the broccoli."

Tomorrow, I'll have her start painting the superstructure. There's a lot of painting on a ship this size. That'll teach her to watch what time it is.

As I climb below, I catch a glimpse of her lights and stop to watch her land. It's typical Cherry. She makes a short, shallow turn, rather than circling and doing it right, and she comes in fast, slapping hard and screeching in the cable, leaving two yards of rubber on the deck. Kids.

I take a deep breath. It's dark now here on the U.S.S. *Fortitude*. The running lights glow in the sea air. The wake brims behind us. As Cherry turns to park on the elevator, I see that her starboard Sidewinder is missing. Sometimes you feel that you're wasting your breath. How many times have we gone over this? If she's old enough to fly, she's old enough to keep track of her missiles. But she's been warned, so it's O.K. by me. We've got plenty of paint. And, as I said, this is a big ship.

## Some final hints on writing about literature

Read the work once carefully.

Decide on a subject generally.

Narrow the subject as much as you can before a second reading.

Ask yourself questions as you read the work a second time, a third, and so on.

Write down the answers to your questions.

Arrange the answers by topics.

Decide on a final approach, topic, title, and thesis.

Discard the erroneous answers to your questions.

Discard the irrelevant points. Choose the one you wish to discuss.

Select the best topics.

Put them in logical order.

Rearrange them as you carry out the creative process of thinking through the subject and writing the paper.

Revise as many times as necessary.

Do not display or emphasize technical terms, which may sometimes be as distracting as they are helpful. Your reader will not be impressed by attempts to show off. That is no substitute for straight thinking and clear writing.

Be sure early in the paper to name the work that you write about and the author (even though you may have this information in the title). Do not misspell the author's name—as often happens. Ernest Hemingway, for example, spelled his last name with only one *m*.

■ **Exercise 1**

*Write down a list of questions about "On the U.S.S. Fortitude." Make a list of topics for papers. Write a paper about "On the U.S.S. Fortitude" before you read the model paper that follows.*

## Model paper

Dealing with the Darkness: Ron Carlson's

"On the U.S.S. Fortitude"

The narrator of Ron Carlson's short story "On the U.S.S. Fortitude" is a woman with a great responsibility. Hers is "a single-parent family," she says—no husband to help her support and guide her two children. The story is a glimpse into her mind that reveals how she sees her responsibility as a mother and how she copes with it, especially how she deals with the threats represented by darkness.

Nighttime, the setting of the story, is greatly on the narrator's mind. She states in the first sentence: "Some nights it gets lonely here on the U.S.S. Fortitude." The nights bother her most because then she feels most keenly her loneliness. The darkness brings more than loneliness, however: it carries the threat of danger to her children. They are still young and highly vulnerable. Though the older child, Dennis, has become confident that he can handle himself in any situation and the daughter, Cherry, is starting to be impatient with her mother's worrying, still they need a strong hand to direct and correct them, to bring them through the darkness, which is emphasized throughout the story.

Through imagination, fortitude, and discipline, the mother copes with the darkness that represents loneliness to her and danger to her children. The most apparent result of her vivid imagination is her fantasy of being the captain of an aircraft carrier. In a metaphor that is both impressively sustained and humorous, she thinks of their new living quarters--no longer cramped as they were at the Hotel Atlantis in the city--as the carrier U.S.S. Fortitude, her children as naval fighter pilots, and herself as the ship's commander plotting the whereabouts of the pilots on radar and awaiting with some anxiety and impatience their return to the ship. In a sense she is playing a game, dreaming. When she was a child and became interested in tennis and its codelike language, her mother called her a dreamer, and so she is. But her dreaming, her active imagination, has provided her with a way of coping with the difficult circumstances in her life. The military metaphor she imaginatively creates works well in the story because every mother fights a kind of war in bringing up her children, battling the forces of darkness to keep them safe and to educate them properly.

The narrator has designated her imaginary carrier the Fortitude, an appropriate name since that is a quality she strongly exhibits. She knows that courage, determination, and endurance--the ingredients of fortitude--are all necessary not only for a military commander but also for a mother trying to raise her children alone. She has much in common with the captain of a naval aircraft carrier. Obviously fortitude is necessary to engage in the military campaigns of war, but it also takes fortitude for a parent to fight off the threats of darkness to her children. "Doing the right thing," says the narrator of "On the U.S.S. Fortitude," is difficult because it "sometimes means getting cursed by your kids."

Nevertheless, she will courageously do the right thing because of her self-discipline and because she realizes the importance of discipline in the lives of her children. "They can love me later," she says, sounding like a military leader who trains complaining troops hard for combat. She will make them follow orders and will punish them when they do not. Discipline, she feels, will help them safely through the darkness. When she was a little girl, she was attracted to the game of tennis because, as she puts it, "I liked the idea of limits." Though she does not try to hold her children back and prevent them from maturing, she stresses the importance of limits, of lines drawn beyond which they should not go. Her rule of life is to play the game well within the lines and to teach her children to do the same thing. That does not mean living without adventure or delight, for she has herself started flying, her way of suggesting that she is taking on new and different ventures.

In "On the U.S.S. Fortitude" Ron Carlson has written a story of hope. Though the mother-narrator has gone through some hard times, stars are now appearing in the night. Her imagination, fortitude, and discipline have proved to be her stays against the darkness. She has developed a sustained comparison between her role and that of a military commander, but throughout the story the captain in her is never more important than the mother. If she charts the course, she also "wipe[s] everything down and sweep[s] the passageways." If she is the commander, she is also, as she makes clear to Cherry, "your mother on the Fortitude."

### ■ Exercise 2

1. *Write a brief description of the methods, techniques, and subject of criticism in the model paper.*
2. *Suggest other topics for critical papers about Carlson's story.*

### ■ Exercise 3

*Write a critical paper about a short story assigned or approved by your instructor.*

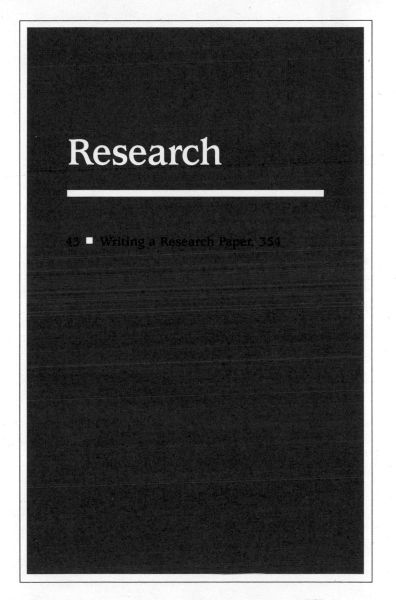

# Research

# 43 Writing a Research Paper

Research is the act of gathering new information. In setting out on a research project, your purpose should not be merely to find enough material to satisfy your instructor that you have used the library and know how to find, quote from, and cite sources. The primary reason for research is to learn enough about a subject to write or speak about it knowledgeably. Research is necessary if you are to know what you are talking about.

After discovering new information and examining and evaluating various ideas, you will arrive at a crucial point: your basic research is finished, but the project is not. A good research paper is not simply a series of paraphrases, citations, and quotations. Formulate your own thesis, give your own slant to the subject, think through questions, and present some conclusions of your own. A research paper embodies the results of exploration into your own thinking as well as into the thinking of others.

**43a** Choose a subject area that interests you, and write on a specific aspect of that subject.

Above all, get started early. In a research project, delay is deadly. Begin by browsing through current periodicals, encyclopedias, indexes (many of which you can consult on computer disks) like the *New York Times Index,* and bibliographic guides that a reference librarian can point out to you. Do not browse too long, however, or in a too-leisurely way. You can easily waste precious time if you do not remind yourself repeatedly that at this early stage, you are doing research; you are searching for a good topic, not merely reading and looking randomly.

## *Choose a general subject area*

Begin the process of selecting a topic by asking yourself what truly interests you. Perhaps you would like to learn more about photography, the fast-food industry, or the Roman Empire. The possibilities are infinite. Free your imagination and let it roam. If nothing comes to mind, and browsing in magazines and a good encyclopedia produces no positive results, turn to a list of broad areas such as the following, decide on one you like, and then go on to narrow it down.

| | | | |
|---|---|---|---|
| advertising | acting | law | television |
| religion | literature | art | sports |
| ecology | philosophy | history | government |
| science | anthropology | economics | medicine |
| marketing | industry | sociology | archaeology |
| psychology | folklore | agriculture | education |

## *Get more specific*

Suppose you have chosen medicine for your general area. Naturally, you will need to focus on one aspect of this vast field. To help yourself decide on an aspect of the area you have selected on the list, draw a circle around it and then review the other entries on the list or add more topics that represent your secondary interests. Draw arrows from your primary area to the secondary ones, as in the example below:

| | | | |
|---|---|---|---|
| advertising | acting | law | television |
| religion | literature | art | sports |
| ecology | philosophy | history | government |
| science | anthropology | economics | (medicine) |
| marketing | industry | sociology | archaeology |
| psychology | folklore | agriculture | education |

By associating a major area of interest with secondary (but vital) concerns, you move closer to a subject. For instance, think of aspects of medicine that involve law and sociology. After more thinking and general reading, a specific topic may emerge, such as "The Effect of Malpractice Suits on the Public's Image of the Physician."

With an entirely different set of interests, you might circle, say, *science* and then draw arrows to *history, geography, anthropology, archaeology,* and *ecology.* Many possibilities for a good topic exist in such a combination, but this procedure brings to light a pattern of interest that gets you closer to your final subject (which might be in this instance "What Happened to the Dinosaurs: A Continuing Debate"). The author of the model paper on pp. 401–453 circled *marketing* and then drew arrows to *psychology* and *advertising.* Some hard thinking then resulted in a good subject, "Illusions in the Mail: Magazine Sweepstakes."

## Avoid inappropriate subjects

After you have tentatively selected a subject, trim it, refine it, and polish it into a definite title and then ask yourself the following questions about it.

1. *Is it too broad?* If you set out to write a research paper of two thousand words on "Religions of the World," the paper will probably be too general, undetailed, and shallow. You might be very much interested in "East Africa: 1824–1886," but this is a subject for a book. The broad scope of such titles is a clue that you will need to narrow your focus to find a workable subject.

2. *Is it too technical?* Consider this question from two angles: yours and your audience's. You may know a great deal about "Bronchial Morphometry in Odontocete Cetaceans," but an audience of general readers will not. Try to reconstitute your subject into one that will be more understandable.

3. *Does it promise more than it can deliver?* Examine your subject carefully to be sure that the promise implicit in the title can be fulfilled. For example, how could a research paper live up to the expectations created by a subject such as "Is There a God?"

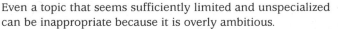
Even a topic that seems sufficiently limited and unspecialized can be inappropriate because it is overly ambitious.

4. *Is it too close to home?* Sometimes controversial issues create such an emotional response in the writer that a sustained, even tone is impossible. If volatile subjects such as abortion, television ministers, racism, and government-sponsored lotteries cause you such anguish that you cannot be calm and objective, avoid them. Interest is one thing; subjective indulgence is another. Controversial issues can be excellent and compelling topics for papers, but be certain that you are not so deeply involved that you will get carried away emotionally.

5. *Is it dull?* You must stand outside yourself to answer this question. You may be fascinated with "Four Types of Bathroom Tubs," but ask yourself if others will be. Usually you can relate what might be an uninteresting topic to other concerns and produce a promising subject. If for some reason you are intrigued with bathtubs, you might develop an imaginative subject by linking that interest with art ("The Bathtub in Postmodern Impressionism") or with psychology ("Psychological Implications of Modern Bathtub Designs").

## 43b Follow a search strategy for using the library.

Once you have determined your subject, you should compile a working bibliography, a list of publications pertinent to your topic. The items on this list should include only the author's name, the title, and any information you need to find the source in the library. Where to begin and how to proceed in your search are matters that require planning; otherwise, you can waste countless hours in random and fruitless excursions. Usually the best place to begin is the **reference room**. Your teachers in this area are the reference librarians, indispensable professionals in any college or university library. They can save you time by suggesting shortcuts or other alterations in your search strategy; they can also direct you to helpful guides, such as the *Library of Congress Subject*

*Headings,* which enables you to find references under subject headings in the library catalog.

## General reference aids

Before you turn to the library's catalog, however, look up your subject in a general or specialized encyclopedia, dictionary, or other such reference works. This is often an excellent starting point because the entries are usually general enough to give you a broad outline of information.

Sometimes articles in encyclopedias conclude with suggestions for further reading (a **bibliography**), which can start you on your way toward compiling a list of materials to read or consult. Below is a listing of encyclopedias, dictionaries, bibliographies, and other reference aids that you will find useful in the early stages of your research. Ask your reference librarian about additional sources.

GENERAL REFERENCE AIDS

*Articles on American Literature.* 1900–1950; 1950–1967; 1968–1975.

*Cambridge Bibliography of English Literature.* 1941–1957. 5 vols.; *New Cambridge Bibliography of English Literature,* 1969–1977. 5 vols.

*Cambridge Biographical Dictionary.* 1990.

*Cambridge Histories.* Ancient, 12 vols., rev. ed. in progress; Medieval, 8 vols.; Modern, 13 vols.; *New Cambridge Modern History.* 14 vols.

*Chambers World Gazetteer.* 1988.

*Collier's Encyclopedia.*

*Columbia Literary History of the United States.* 1988.

*Contemporary Authors.* 1962–.

*Contemporary Literary Criticism.* 1973–.

*Current Biography.* 1940–. "Who's News and Why."

*Dictionary of American Biography.* 1928–1937. 20 vols. 8 supplements, 1941–1990.

*Dictionary of American History.* Rev. ed., 1976–1978. 8 vols.

*Dictionary of Literary Biography.* 1978–.

*Dictionary of National Biography.* 1885–1901. 22 vols. 10 supplements, 1901–1985.

*Encyclopedia Americana.* Supplemented by the *Americana Annual.* 1923–.

*Encyclopedia Judaica.* 1972. 16 vols. Supplemented by its *Yearbook.*

*Encyclopedia of Philosophy.* 1967. 8 vols.

*Encyclopedia of Religion.* 1987. 15 vols.

*Encyclopedia of World Art.* 1959–1968. 15 vols. 2 supplements, 1983–1987.

*Encyclopedia of World Cultures.* 1991. 6 vols.

*Encyclopedia of World Facts and Dates.* 1993.

*English Novel Explication.* 1973. 4 supplements, 1976–1990.

*Essay and General Literature Index.* 1900–. "An index to Volumes of Collections of Essays and Miscellaneous Works."

*Facts on File: A Weekly World News Digest . . .* 1940–.

*Facts on File: Bibliography of American Fiction.* 1991–.

*Halliwell's Filmgoer's and Video Viewer's Companion.* 10th ed., 1993.

*Harvard Encyclopedia of American Ethnic Groups.* 1980.

*Hutchinson Dictionary of World History.* 1993.

*Information Please Almanac.* 1947–.

*International Encyclopedia of Social Sciences.* 1968. 17 vols. *Biographical Supplement.* 1979. *Social Science Quotations.* 1991.

*Magill's Bibliography of Literary Criticism.* 1979. 4 vols.

*McGraw-Hill Encyclopedia of Science and Technology.* 7th ed., 1992. 20 vols. Supplemented by *McGraw-Hill Yearbook of Science and Technology.*

*McGraw-Hill Encyclopedia of World Drama.* 2nd ed., 1984. 5 vols.

*MLA International Bibliography of Books and Articles on the Modern Languages and Literatures.* 1919–.

*New Catholic Encyclopedia.* 1967–1989. 18 vols.

*New Encyclopaedia Britannica.* Supplemented by *Britannica Book of the Year.* 1938–.

*New Grove Dictionary of Music and Musicians.* 1980. 20 vols.

*New Princeton Encyclopedia of Poetry and Poetics.* 1993.

*Oxford Classical Dictionary.* 2nd ed., 1970.

*Oxford Companion to American Literature.* 5th ed., 1983.

*Oxford Companion to English Literature.* 5th ed., 1985.

*Oxford History of English Literature.* 1945–.

*Statesman's Yearbook: Statistical and Historical Annual of the States of the World.* 1864–.

*Statistical Abstract of the United States.* 1878–.

*Webster's New Geographical Dictionary.* 1984.

*World Almanac and Book of Facts.* 1868–.

### The library catalog

The basic tool for finding books in a library is the catalog. It may be either a card catalog or a computerized catalog. Books are listed by author, by title, and by subjects, with helpful cross-references that will steer you to new aspects of your topic. Traditional card catalogs have library holdings recorded on three-by-five-inch cards that are filed alphabetically by author, by title, and by subject. Reproduced on pp. 361–362 are three typical catalog cards—actually three copies of the same library card to be filed in three different ways.

Author Card

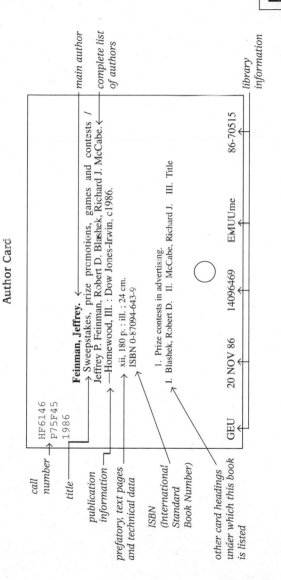

*main author*

*complete list of authors*

*library information*

call number →

title →

publication information →

prefatory; text pages and technical data →

ISBN (International Standard Book Number) →

other card headings under which this book is listed →

HF6146
P75F45
1986

**Feinman, Jeffrey.** ←
→ Sweepstakes, prize promotions, games and contests /
Jeffrey P. Feinman, Robert D. Blashek, Richard J. McCabe. ←
—Homewood, Ill. : Dow Jones-Irwin, c1986.
→ xii, 180 p. : ill. ; 24 cm.
→ ISBN 0-87094-643-9

  1. Prize contests in advertising.
I. Blashek, Robert D.   II. McCabe, Richard J.   III. Title

GEU          20 NOV 86      14096469      EMUUme      86-70515

## Title Card

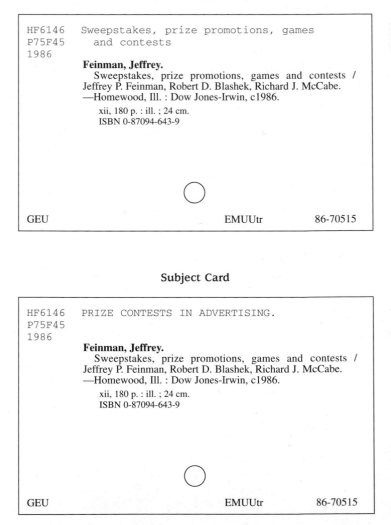

```
HF6146    Sweepstakes, prize promotions, games
P75F45       and contests
1986
          Feinman, Jeffrey.
            Sweepstakes, prize promotions, games and contests /
          Jeffrey P. Feinman, Robert D. Blashek, Richard J. McCabe.
          —Homewood, Ill. : Dow Jones-Irwin, c1986.
              xii, 180 p. : ill. ; 24 cm.
              ISBN 0-87094-643-9




GEU                              EMUUtr              86-70515
```

## Subject Card

```
HF6146    PRIZE CONTESTS IN ADVERTISING.
P75F45
1986
          Feinman, Jeffrey.
            Sweepstakes, prize promotions, games and contests /
          Jeffrey P. Feinman, Robert D. Blashek, Richard J. McCabe.
          —Homewood, Ill. : Dow Jones-Irwin, c1986.
              xii, 180 p. : ill. ; 24 cm.
              ISBN 0-87094-643-9




GEU                              EMUUtr              86-70515
```

Computerized or online catalogs are being used by an increasing number of libraries to take advantage of the computer's capabilities to store and retrieve the information formerly found on catalog cards. These systems allow you to sit at a terminal and search for material by author, title, subject, or any combination of these. Many online catalogs also allow you to search for library holdings in a variety of new ways, such as by publisher, year of publication, library classification number, or International Standard Book Number (ISBN). Often computer modems at locations away from the library provide access to these catalogs.

Online catalogs vary considerably from library to library, but most of the systems are easy to use. They provide many on-screen prompts to guide novice users (see sample below).

## Online Catalog Screen Displays

```
To pick a new button, press TAB or button's first letter.
To begin using the library system, press RETURN or ENTER now.

HELP        PRINT
BEGIN       END         REQUEST
```

```
                    UNICORN COLLECTION MANAGEMENT SYSTEM

                           WELCOME TO EUCLID

    Emory University Computing and Library Information Delivery System

             Above the line are buttons such as HELP and BEGIN

    Simply select a button using the TAB key, then press RETURN or ENTER

                 Or just type the first letter of a button
```

```
To pick a new button, press TAB or button's first letter.
Enter the number of the lookup you want to do, then press RETURN or ENTER.

HELP         GOBACK          STARTOVER       PRINT
REQUEST      CHOOSE:1
```

```
                    LOOKUP IN CATALOG BY:

                1) WORDS OR PHRASE

                2) AUTHOR

                3) TITLE

                4) SUBJECT

                5) AUTHOR WITH TITLE

                6) OTHER COMBINATIONS

                7) BROWSING
```

The full records from online catalogs generally include the same information that is found on cards in a card catalog. Below is a sample online catalog entry for a book on sweepstakes.

## Online Catalog Entry

Online catalog
Authors, editors, etc.
Full Information   Document   30940966

| Title: | Sweepstakes, prize promotions, games and contests |
| Publisher: | Dow Jones-Irwin, Homewood, Ill., 1986 |

| Author(s): | Feinman, Jeffrey. / Blashek, Robert D. / McCabe, Richard J. |
| Subject: | Prize contests in advertising. |

Contains: xii, 180 p. : Ill.; 24 cm.
ISBN 0870946439 : $19.95

| Copies: | Location | Call number | Status | Due |
|---------|----------|-------------|--------|-----|
| General | wg | HF6146.P75 F45 1986 | Available | |

NOTE: In your research, you may need to consult *both* the card catalog and the computerized catalog. Check with a reference librarian.

### Periodical indexes

Articles in magazines and newspapers are a rich source of information on almost every subject. Some important periodical indexes are listed below. Some are in book format, some are searchable by computer as databases or on compact disk (CD or CD-ROM), some are on microfilm or microfiche, and some are available in more than one format. Usually the book versions cover more years than the computerized versions, especially on a topic that is not a current one. The *Readers' Guide to Periodical Literature,* for example, found in many libraries in large green volumes in the Reference Department, goes back to 1900. It is also available on compact disk or as a database to be used online going back to 1983. Your reference librarian can tell you which indexes are available in what format, how far back they go, and how you can use them.

PERIODICAL INDEXES

Titles preceded by an asterisk (*) are available in online or CD versions. Dates of coverage refer to the printed versions.

*ABI/INFORM*. 1971–.
Covers periodicals in the business and management fields.

*Applied Science and Technology Index*. 1958–.
Subject index to periodicals in the fields of aeronautics, automation, chemistry, electricity, engineering, physics, and other areas.

*Art Index*. 1929–.
Author and subject index to selected fine arts periodicals.

*Biography Index*. 1946.
Indexes biographical material appearing in books and magazines.

*Biological and Agricultural Index.* 1964–.
> Continues *Agricultural Index.* 1919–1964. "Cumulative subject index to periodicals in the fields of biology, agriculture, and related sciences."

*Book Review Digest.* 1905–.
> An index to book reviews, including excerpts from the reviews.

*Book Review Index.* 1965–.
> Indexes a large number of general-interest and scholarly periodicals.

*British Humanities Index.* 1962–.
> Supersedes in part *Subject Index to Periodicals,* 1915–1922, 1926–1961. Subject index to British periodicals.

*Business Periodicals Index.* 1958–.
> Cumulative subject index to periodicals in all fields of business and industry.

*Current Index to Journals in Education.* 1969–.
> Covers "the core periodical literature in the field of education" and "peripheral literature relating to the field of education."

*Education Index.* 1929–.
> "A Cumulative Author and Subject Index to a Selected List of Educational Periodicals and Yearbooks."

*General Science Index.* 1978–.
> Science literature for the nonspecialist.

*Humanities Index.* 1974–.
> Preceded by *Social Sciences and Humanities Index.*

*Index to Black Periodicals.* 1950–.
> "An Index to Afro-American Periodicals of General and Scholarly Interest."

*Industrial Arts Index.* 1913–1957.
> "Subject Index to a Selected List of Engineering, Trade and Business Periodicals." Became *Applied Science and Technology Index* and *Business Periodicals Index,* above.

*Infotrac.* 1980–.

A compact disk system available in a variety of editions. Indexes newspapers and general-interest, academic, and business periodicals.

*Music Index.* 1949–.

An author and subject index to a comprehensive list of music periodicals published throughout the world.

*National Newspaper Index.* 1979– .

Indexes the *New York Times, Wall Street Journal, Christian Science Monitor, Los Angeles Times,* and *Washington Post.*

*New York Times Index.* 1851–.

"Master-Key to the News Since 1851."

*Nineteenth Century Readers' Guide to Periodical Literature.* 1890–1900.

Author and subject index to some fifty English-language general periodicals of the last decade of the nineteenth century.

*PAIS. . . .* 1915–.

The Public Affairs Information Service index to periodicals and government publications chiefly in the social sciences.

*Poole's Index to Periodical Literature.* 1802–1906.

An index by subject to the leading British and American periodicals of the nineteenth century.

*Readers' Guide to Periodical Literature.* 1900–.

An index to the most widely circulated American periodicals.

*Social Science Index.* 1974–.

Preceded by Social Sciences and Humanities Index.

*Social Sciences and Humanities Index.* 1965–1974.

Formerly *International Index.* 1907–1965. Author and subject index to a selection of scholarly journals.

Suppose you are writing about magazine sweepstakes. Looking under **Sweepstakes** in the *Readers' Guide to Periodical Literature,* 1991, you find the following entry:

subject
heading            explanation
               of contents

⎣ ⟶ **Sweepstakes**       ↓

title ⟶ Lucky numbers [rise of entries in direct mail sweepstakes]

author ⟶ J. A. Tooley, il *U.S. News & World Report* 110:12
     Ja 21 '91
      ↑        ↑   ↗ ↖
      date    periodical  volume page

Consult indexes to periodicals in specific subject areas, such as the *General Science Index,* as well as the more general indexes such as *Readers' Guide.* Searching the CD (compact disk) index *ABI/INFORM* under "Magazines and Sweepstakes," you discover the following:

```
Access no:    00642030   ProQuest   ABI/INFORM
Title:        While Most Consumers Think That All Sweepstakes Are
              the Same, Direct Mail Practitioners Know Better.  See
              How Publishers Clearing House Compares to American
              Family Publishers and Vice Versa.
Authors:      Rosenfield, James R.
Journal:      Direct Marketing (DIM)   ISSN: 0012-3188
              Vol: 55  Iss: 6  Date: Oct 1992  p: 38-40  Illus: Charts
Companies:    Publishers Clearing House
              American Family Publishers
Subjects:     Direct mail advertising; Sweepstakes; Manipulation; Tactics
Geo Places:   US
Codes:        7200 (Advertising); 9190 (United States)
Abstract:     The 3 major themes in the sweepstakes mailings of
  Publishers Clearing House (PCH) and American Family Publishers (AFP)
  are choice, loss, and justice.  Sweepstakes emphasize being chosen and
  making a choice.  Both companies' mailings tell customers that they
  are one of the few chosen to participate in the sweepstakes offer.
  Then, the customers receive a clear message that choosing not to
  enter leads to loss.  Both AFP and PCH tell customers that someone
  else will win their prize if they fail to return the winning number.
  There is also the possibility of a future loss by being dropped from
  the mailing list.  This is connected to the theme of justice.  Cus-
  tomers are allowed to enter a sweepstakes without buying anything.
  Yet, psychologically, customers feel that they cannot get something
  for nothing.  PCH and AFP sell magazines by playing on these feel-
  ings.  Telling customers that failure to purchase will result in
  their names being dropped from the mailing list is also persuasive.
```

Notice that the bibliographical entry giving pertinent information about the title, author, place of publication, and subject is accompanied in the above entry by an abstract, a summary of the article. Abstracts are not always available in periodical indexes, but when they are, you can quickly determine whether the article will be of use in your research. Consult, in addition, abstracting services such as the following:

*Academic Abstracts*
*Biological Abstracts*
*Chemical Abstracts*
*Dissertation Abstracts International*
*Psychological Abstracts*
*Sociological Abstracts*

**43c** Distinguish between primary and secondary sources and evaluate materials.

**Primary sources** are those about which other materials are written. *Moby-Dick* is a primary source; writings about *Moby-Dick* are **secondary sources**. The terms distinguish between which material comes first (primary) and which follows (secondary). Sometimes what was once secondary—say, a book about Puritan New England—becomes primary material when it is the subject of later writings (in this instance, articles about the book). Primary sources for a study of tourists would consist of published and unpublished diaries, journals, and letters by tourists, interviews of tourists, and the like. The writings of journalists and historians about tourists would be secondary sources.

When considering which secondary sources to use in your research paper, you must know when a work was written, what information was available to its author, the reputation and reliability of the author, and, insofar as you can judge, the soundness of the author's argument and use of evidence. *The fact of publication is not a guarantee of truth, wisdom, objectivity, or perception.* Look with a critical eye at all the secondary material you discover in your search and rely only on those sources that are worthy (see pp. 249–250).

**43d** Take accurate and effective notes from sources.

After you have compiled a working bibliography (see **43b**), located some of the sources you wish to use, and done some preliminary reading, you are ready to begin collecting specific material for your paper. If you are writing a formally documented paper, make a **bibliography card** for each item as you examine it. This will be a full and exact record of bibliographical information, preferably on a three-by-five-inch filing card. From these cards you will later compile your list of Works Cited or References. A sample card is shown on p. 371. The essential information includes the name of the author, the title of the work, the place and date of publication, and the name of the publisher. If the work has an editor or a translator, is in more than one volume, or is part of a series, these facts should be included. For later checking, record the library call number.

CAUTION:  Carelessness in jotting down call numbers can lead to wasted time and confusion. Double-check to be sure you have recorded them correctly.

## Bibliography Card

*(reduced facsimile—actual size 3 x 5 inches)*

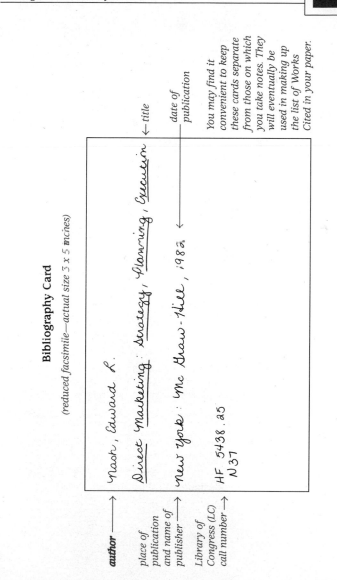

author →    Nash, Edward L.

place of
publication
and name of →    Direct Marketing: Strategy, Planning, Execution    ← title
publisher
New York: McGraw-Hill, 1982 ←    date of
publication

Library of
Congress (LC) →    HF 5438.25
call number    N37

*You may find it
convenient to keep
these cards separate
from those on which
you take notes. They
will eventually be
used in making up
the list of Works
Cited in your paper.*

For **note-taking,** your next step, use cards or slips of paper uniform in size. Cards are easier to use than slips because they withstand more handling. Develop the knack of skimming a source so that you can move quickly over irrelevant material and concentrate on pertinent information. Use the table of contents, the section headings, and the index to find chapters or pages of particular use to you. As you read and take notes, consider what subtopics you will use. The two processes work together: your reading will give you ideas for subtopics, and the subtopics will give direction to your note-taking.

At this point, you are already in the process of organizing and outlining the paper. Suppose you wish to make a study of direct-mail magazine sweepstakes. You might work up the following list of tentative subjects:

history of magazine-subscriptions selling

role of computers in promoting sweepstakes

the gambling urge

deceptiveness of magazine sweepstakes

hope and diversion through sweepstakes participation

designers of sweepstakes as illusionists

You may not stick to the order in which you list such headings; in fact, you may even rework or replace the topics themselves as you go along. A successful research project is a process in which your thought changes and develops as you read, ponder, and take notes.

To illustrate the methods of note-taking, suppose you have found the following passage about the "response device" of magazine sweepstakes materials. (The authors in this passage are addressing sweepstakes designers and promoters, not the public.)

The response device is both your entry form and your order form. Generally, in direct mail your aim is to close the sale right then and there. Because of this, you want to compel your recipients not only to enter the sweepstakes but to say yes to your product as well. Because you cannot require purchase with a sweepstakes entry, many people enter the sweepstakes without a thought of buying. This is where the yes/no psychology comes in to play.

Somewhere in your direct mail package you will want an involvement device that encourages prospects to buy your product and discourages them from just entering without buying. This device can be a perforated token or a peel-off sticker that can be affixed to the outer envelope or the response device. This is where yes/no stickers come in. Psychological research tells us that people really like to be good and please others. In essence, they like saying yes and they don't feel good when they have to say no. (In fact, in the 1970s there was a successful pop psychology book called *When I Say No I Feel Guilty*.)

JEFFREY P. FEINMAN, ROBERT D. BLASHEK, AND RICHARD J. McCABE
*Sweepstakes, Prize Promotions, Games and Contests*

You may make notes on this passage by paraphrasing, by quoting, or by combining short quotations with paraphrasing (see 43f). The card on p. 374 identifies the source, gives a subject heading, indicates the page number, and then records relevant information in the student's own words. It *extracts* items of information instead of merely recasting the entire passage and line of thought in different words. Notice the careful selection of details and the fact that the paraphrase is considerably shorter than the original. The first card on p. 375 records a direct quotation, and the second combines quotation with paraphrase.

If you cannot determine just what information you wish to extract when you are taking notes, copy the entire passage. For later reference you must be careful to show by quotation marks that it is copied verbatim.

NOTE: The writer inserted a slash [/] in the quotation on the first card on p. 375 to indicate a break between pages.

## Paraphrased Notes

*(reduced facsimile—actual size 3 x 5 inches)*

subject
heading →

→ Sweepstakes and Psychology

Leinman, Blaschek, and McCabe ←—— Identification of source.
Full bibliographical
information has been
taken down on the
bibliography card.

page
number → 184-85  A knowledge of human psychology
is essential to those who sell through
magazine sweepstakes. Promoters are aware
that most people are more comfortable
saying "yes" than "no"; therefore, sweepstakes
materials include a "yes/no" box with
instructions that entrants must affix a
sticker or otherwise indicate to the place that
signifies whether they are buying or not. Marketers
count on the natural human tendency to
prefer "yes."

## Quotation

Sweepstakes and Psychology

Feinman, Blashek, and McCabe

124-25 "Psychological research tells us
that people really like to be good and
please others. In essence, / they
like saying yes and they don't feel
good when they have to say no."

## Quotation and Paraphrase

Sweepstakes and Psychology

Feinman, Blashek, and McCabe

124-25 "Psychological research tells us
that people really like to be good and
please others. In essence, / they like
saying yes and don't feel good when they
have to say no." Therefore, sweepstakes
materials include a "yes/no" box with
instructions that entrants must affix
a sticker or other object to the place
that signifies whether they are buying
or not.

## Quotation

> Appeal of Greed
>
> Rosenspan
>
> 367 "In the last four years, sheer, naked greed has been as effective a selling tool as anything else you can name. Publishers Clearing House and American Family Publishers have practically co-opted this one. In fact, you have to read their packages very carefully to find out that they're actually selling magazine subscriptions."

## Quotation

> Success of Sweepstakes
>
> Rosenspan
>
> 367 "They are certainly successful. After all, how many businesses could afford to give away ten million dollars year after year after year?"

## Paraphrase

> ### Appeal of Greed
>
> Rosenspan
>
> 367   Selling flourishes when it appeals to greed. In recent times, major magazine subscription companies have been preeminent in proving this point with the success of their sweepstakes, which play strongly to greed and disguise their main purpose of selling.

## Quotation and Paraphrase

> ### Success of Sweepstakes
>
> Rosenspan
>
> 367   If anyone doubts the financial effectiveness of sweepstakes, it is necessary only to look at the success of such firms as American Family Publishers, for "how many businesses could afford to give away ten million dollars year after year after year?"

A note card should contain information from only one source. If you keep your notes on cards with subject headings, you can later arrange all your cards by topics—a practical and an orderly procedure. Remember, the accuracy of your paper depends largely on the accuracy of your notes.

A photocopying machine can guarantee accuracy and save you time. At an early stage in your research, it is not always possible to know exactly what information you need. Photocopy some of the longest passages, and then you can study and digest them and take notes later.

Read the following passage, which deals with greed as a factor in the creation of magazine sweepstakes materials.

> Whether or not multimillion-dollar state lotteries have helped develop and nurture this sin, it seems to be more popular than ever before. In the last few years, sheer, naked greed has been as effective a selling tool as anything else you can name. Publishers' Clearing House and American Family Publishers have practically co-opted this one. In fact, you have to read their packages very carefully to find out that they're actually selling magazine subscriptions. But it is my belief that these highly professional packages would work no matter what product or service they represented. They are certainly successful. After all, how many businesses could afford to give away ten million dollars year after year after year?
>
> ALAN ROSENSPAN
> "Psychological Appeals"

Now study the note cards on pp. 376–377, which are based on this passage. Observe the difference in subject headings and treatments.

## 43e Produce an effective outline.

Because a research paper is usually longer and more complex than ordinary college papers, a good outline is likely to be of even greater help in your attempts to organize, expand, and delete materials. Consult 41c about the various kinds of outlines.

After you have worked out a tentative outline, study it to be sure that you have included all the areas you wish to cover, that you have grouped subjects under the most appropriate headings, and that you have arranged the aspects of your discussion in the order that will produce the maximum effectiveness.

The thought that you put into a full and coherent outline is an excellent investment, which will pay off when you write the paper. You will find that you have worked out many of the problems of organization and made essential connections in preparing the outline that you would otherwise have to face during the composition of the paper. See the model outline on pp. 291-292.

**43f**  Acknowledge your sources; avoid plagiarism. Quote and paraphrase accurately.

Using others' words and ideas as if they were your own is a form of stealing. It is called **plagiarism.** Avoid it by giving full references to sources.

**All direct quotations must be set off or placed in quotation marks and acknowledged in your text.**

Even when you take only a phrase or single distinctive word from a source, you should enclose it in quotation marks and indicate where you got it. Compare the sources below with the passages that plagiarize them.

SOURCE

> More than any previous explorer, Cook was well prepared to chart his discoveries and fix their locations accurately. He sailed at a time of rapid advances in methods of navigation; his ship was equipped with every available type of scientific instrument; he had the services of professional astronomers; and he himself had a far more sophisticated understanding of astronomy, mathematics, and surveying techniques than most ship captains.
>
> LYNNE WITHEY
> *Voyages of Discovery: Captain Cook and the Exploration of the Pacific*

PASSAGE WITH PLAGIARISM

> Although we often think of the famous Captain Cook as someone who was more of an adverturer than a systematic explorer, he actually was better prepared **than any previous explorer** to discover and fix new **locations accurately.** Many improvements were taking place in sailing. For example, **his ship was** furnished out **with every available type of scientific instrument,** and Cook **himself had a far more sophisticated understanding** of navigation than is often realized.

SOURCE

> Although Poe's protagonists question their thoughts and feelings as well as their motivations and analyze the amount of harm they have done, they rarely grow in understanding. They lack the detachment necessary to rectify the situation and the will to carry out their decisions once these are made. Solipsistic in every way, they are caught in the stifling world of their own choice.

> BETTINA L. KNAPP<br>*Edgar Allan Poe*

PASSAGE WITH PLAGIARISM

> Poe's main characters do examine themselves, but **they rarely grow in understanding.** They are not objective enough to change, and they lack **the will to carry out their decisions once these are made.** They are weird, or **solipsistic, in every way** and therefore get stifled in **their own choice.**

These passages with plagiarism do contain some paraphrasing, but they also use some of the words of their sources without employing quotation marks, and neither indicates a source. If the source were acknowledged at the end but no quotation marks were used around words and phrases taken from the source, the writer would still be guilty of plagiarism. The writer of the passage about Poe was not only dishonest but also careless. He or she obviously did not know the meaning of the word *solipsistic* in the source, guessed that it meant something like "weird," and thus composed a sentence that does not make sense. The writer assumed that the reader would not know what *solipsistic* means either and would not have read the source. Thus plagiarism is also insulting to the reader and impractical since it often calls attention to itself.

## Good and bad paraphrasing

A good paraphrase expresses the ideas found in the source (for which credit is always given) but not in the same words. It preserves the sense, but not the form, of the original. It does not retain the sentence patterns and merely substitute synonyms for the original words, nor does it retain the original words and merely alter the sentence patterns. It is a genuine restatement. It is briefer than its source.

SOURCE

Although Poe's protagonists question their thoughts and feelings as well as their motivations and analyze the amount of harm they have done, they rarely grow in understanding. They lack the detachment necessary to rectify the situation and the will to carry out their decisions once these are made. Solipsistic in every way, they are caught in the stifling world of their own choice.

IMPROPER PARAPHRASE

Even though Poe's heroes ask themselves about their ideas and emotions and also their reasons for acting and figure out how much damage they have left behind, they do not often gain more wisdom. They do not have the impartiality mandatory to alter the condition and the determination to follow through on their resolutions when they are formulated. Self-involved in all manner, they are snared in the smothering realm of their making (Knapp 155).

PROPER PARAPHRASE

Poe's main characters do engage in self-examination, but they seldom benefit from it. They are too close to their own problems to solve them, too weak to change. Their whole world is within themselves, and they are destroyed by it (Knapp 155).

Frequently a combination of quotation and paraphrase is effective.

SOURCE

> More than any previous explorer, Cook was well prepared to chart his discoveries and fix their locations accurately. He sailed at a time of rapid advances in methods of navigation; his ship was equipped with every available type of scientific instrument; he had the services of professional astronomers; and he himself had a far more sophisticated understanding of astronomy, mathematics, and surveying techniques than most ship captains.

QUOTATION AND PARAPHRASE

> One of Captain Cook's distinctions was the care he took in readying himself for his voyages. He "was well prepared to chart his discoveries and fix their locations accurately." He did not sail without good equipment and competent navigators, and "he himself had a far more sophisticated understanding of astronomy, mathematics, and surveying techniques than most sea captains" (Withey 87–88).

PARAPHRASE

> One of Captain Cook's distinctions was the care he took in readying himself for his voyages. He did not sail without good equipment and competent navigators. Indeed, he was well versed in the several sciences so important in long-distance sailing (Withey 87–88).

## ■ Exercise 1

*Choose an article from a periodical; avoid articles that are mainly factual or statistical. Then select two passages from one page of the article and make a photocopy (number the passages 1 and 2). In ten to fifteen words, paraphrase the first passage (which should be thirty to fifty words long). Do not state any ideas of your own. Do not use words from the article. In about twenty-five words, paraphrase the second passage (which should be sixty to seventy-five words long). This time combine some of your own ideas with the paraphrase of the source. Turn the photocopy in with your paraphrases.*

■ **Exercise 2**

*Turn in to your teacher a photocopy of a single page from an article in a periodical. (Do not use the same page that you used for Exercise 1.) Draw circles around three quotations. Submit with the photocopy the following: (a) a sentence that you quote with no writing of your own, (b) a sentence of your own with short quotations in it from the source, and (c) a sentence of your own that introduces a quoted sentence. Write your part, followed by a comma or a colon, and then the complete quoted sentence.*

**43g** Follow an accepted system of documentation.

The forms of documentation vary with fields, periodicals, and publishers. Instructors frequently have preferences that determine what specific system you will need to conform to in your research paper. The following style books and manuals are often used in general studies.

Achtert, Walter S., and Joseph Gibaldi. *The MLA Style Manual.* New York: Modern Language Assn. of America, 1985.

*Associated Press Stylebook and Libel Manual.* Reading: Addison, 1992.

*Chicago Manual of Style.* 14th ed. Chicago: U of Chicago P, 1993

Gibaldi, Joseph. *MLA Handbook for Writers of Research Papers.* 4th ed. New York: Modern Language Assn. of America, 1995.

*Publication Manual of the American Psychological Association.* 4th ed. Washington: American Psychological Assn., 1994.

Rubens, Philip, ed. *Science and Technical Writing: A Manual of Style.* New York: Holt, 1992.

Skillin, Marjorie E., Robert M. Gay, et al. *Words into Type*. 3rd ed. Englewood Cliffs: Prentice, 1974.

Turabian, Kate L. *A Manual for Writers of Term Papers, Theses, and Dissertations*. 5th ed. Chicago: U of Chicago P, 1987.

United States Government Printing Office. *Style Manual*. Washington: GPO, 1984.

*Webster's Standard American Style Manual*. Springfield, MA: Merriam-Webster, 1985.

The model paper on pp. 401–453 follows the system of documentation described in *The MLA Handbook*. This system is widely used in the humanities.

The model paper on pp. 463–485 illustrates the method of documentation presented in the *Publication Manual of the American Psychological Association* (APA), most frequently used in the social sciences.

# 43h The MLA Style of Documentation

## List of works cited

At the end of a research paper, after the text and after the notes (or endnotes), comes a section called "Works Cited." Here you list alphabetically only the sources that you refer to in the text (see the list for the model paper, pp. 449–453). Study the examples below of forms used in MLA documentation for various kinds of sources.

### A BOOK BY A SINGLE AUTHOR

Bell, Bernard W. <u>The Afro-American Novel and Its Tradition</u>. Amherst: U of Massachusetts P, 1987.

Marcus, Millicent. <u>Italian Film in the Light of Neorealism</u>. Princeton: Princeton UP, 1986.

### A PAMPHLET

Toops, Connie M. <u>The Alligator: Monarch of the Everglades</u>. Homestead, FL: Everglades Natural History Association, 1979.

NOTE: A pamphlet is treated as a book. Name the state when the place of publication (Homestead) is not likely to be known. Use U.S. Postal Service abbreviations.

### AN EDITED BOOK

Murry, John Middleton, ed. <u>Journal of Katherine Mansfield</u>. New York: Knopf, 1952.

**A BOOK WITH TWO EDITORS OR AUTHORS**

Manley-Casimir, Michael, and Carmen Luke, eds.
<u>Children and Television: A Challenge for
Education</u>. New York: Praeger, 1987.

NOTE: On the title page of this book the following cities are listed as places of publication: New York; Westport, Connecticut; and London. In your entry, use only the first city named.

**A BOOK WITH MORE THAN THREE EDITORS OR AUTHORS**

Kagan, Sharon L., et al., eds. <u>America's Family
Support Programs</u>. New Haven: Yale UP, 1987.

NOTE: The title page of this book lists four editors: Sharon L. Kagan, Douglas R. Powell, Bernice Weissbourd, and Edward F. Zigler. Use only the first, followed by *et al.* ("and others").

**A TRANSLATED BOOK**

Palomino, Antonio. <u>Lives of the Eminent Spanish
Painters and Sculptors</u>. Trans. Nina Ayala
Mallory. Cambridge: Cambridge UP, 1987.

**A MULTIVOLUME WORK**

Hendrick, Burton J. <u>The Life and Letters of Walter
H. Page, 1855 to 1918</u>. 2 vols. Garden City:
Doubleday, 1927.

**A REPUBLISHED BOOK**

Langdon, William Chauncy. <u>Everyday Things in
American Life, 1776-1876</u>. 1941. New York:
Scribner's, 1969.

NOTE: The first date is that of original publication.

SECOND OR LATER EDITION OF A BOOK

Brooks, Cleanth, and Robert Penn Warren. <u>Modern
    Rhetoric</u>. 2nd ed. New York: Harcourt, 1958.


AN ESSAY IN A COLLECTION OF ESSAYS OR AN ANTHOLOGY

Powell, Douglas R. "Day Care as a Family Support
    System." <u>America's Family Support Programs</u>.
    Ed. Sharon L. Kagan, et al. New Haven: Yale UP,
    1987. 115-32.


AN INTRODUCTION, FOREWORD, PREFACE, OR AFTERWORD IN A BOOK

Gibson, James. Introduction. <u>The Complete Poems
    of Thomas Hardy</u>. London: Macmillan, 1976.
    xxxv-xxxvi.

Tobias, Andrew. Foreword. <u>Extraordinary Popular
    Delusions and the Madness of Crowds</u>. By
    Charles Mackay. 1841. New York: Harmony,
    1981. xii-xvi.

NOTE: The word is *foreword*, not *forward*. This reference is to a
modern printing of an older work.


MORE THAN ONE WORK BY SAME AUTHOR

Phillips, Derek, L. <u>Abandoning Method</u>. San
    Francisco: Jossey, 1973.

---. <u>Toward a Just Social Order</u>. Princeton:
    Princeton UP, 1986.

---. <u>Wittgenstein and Scientific Knowledge: A Socio-
    logical Perspective</u>. Totowa, NJ: Rowman, 1977.

NOTE: Multiple works by the same author are listed alphabetically
by title.

**SIGNED ARTICLE IN A REFERENCE WORK**

Emmet, Dorothy. "Ethics." <u>International Encyclopedia of Social Sciences</u>. 1968 ed.

**UNSIGNED ARTICLE IN A REFERENCE WORK**

"Domino." <u>The Encyclopedia Americana</u>. 1986 ed.

**GOVERNMENT PUBLICATION**

United States. Dept. of Commerce. Internal Trade Administration. <u>United States Trade: Performance in 1985 and Outlook</u>. Washington: GPO, 1986.

**ARTICLE IN A MONTHLY PERIODICAL**

Clark, Earl. "Oregon's Covered Bridges." <u>Travel—Holiday</u> Jan. 1987: 71-72.

**ARTICLE IN A WEEKLY PERIODICAL**

Bernstein, Jeremy. "Our Far-Flung Correspondents." <u>New Yorker</u> 14 Dec. 1987: 47-105.

**ARTICLE IN A QUARTERLY PERIODICAL** (with continuous pagination)

Handlin, Oscar. "Libraries and Learning." <u>American Scholar</u> 56 (1987): 205-18.

**ARTICLE IN A JOURNAL (not with continuous pagination)**

Ardoin, John.  "A Pride of Prima Donnas."  Opera
    Quarterly 5.1 (1987): 58-70.

NOTE:  The issue number (1) has been added.

**ARTICLE WITH TWO AUTHORS IN A JOURNAL**

Gottfries, Nils, and Henrik Horn.  "Wage Formation
    and the Persistence of Unemployment."  Economic
    Journal 97 (1987): 877-84.

**BOOK REVIEW IN A PERIODICAL**

Davin, Delia.  Rev. of Revolution Postponed: Women
    in Contemporary China, by Margery Wolf.  Public
    Affairs 59 (1986-87): 684-85.

**SIGNED ARTICLE IN A DAILY NEWSPAPER**

Patton, Scott.  "Rules of the Ratings Game."
    Washington Post 19 Jan. 1988, late ed.: D7.

NOTE:  "D" refers to the section of the paper; "7," to the page
within that section. For newspaper sections designated by
numbers instead of letters, add "sec": sec. 4: 7. For newspapers
not divided into sections, simply give the page number after
the date: 19 Jan. 1988: 7. The edition should be specified when
possible.

**UNSIGNED ARTICLE IN A DAILY NEWSPAPER**

"Space Station Is Falling, to Hit Earth in February."
    Orlando Sentinel 5 Jan. 1991: A16.

**SIGNED EDITORIAL IN A DAILY NEWSPAPER**

Blair, Sharon. "A Charity that Breeds Success."
      Editorial. <u>Wall Street Journal</u> 31 Dec. 1990: 6.

NOTE: No section is indicated before the page number because this particular issue of the newspaper was numbered consecutively throughout.

**UNSIGNED EDITORIAL IN A DAILY NEWSPAPER**

"The Odds of Ignorance." Editorial. <u>Miami Herald</u>
      19 Jan. 1988: A10.

**UNPUBLISHED DISSERTATION**

Kenney, Catherine McGehee. "The World of James
      Thurber: An Anatomy of Confusion." Diss.
      Loyola of Chicago, 1974.

**ABSTRACT IN *DISSERTATION ABSTRACTS INTERNATIONAL***

Kenney, Catherine McGehee. "The World of James
      Thurber: An Anatomy of Confusion." Diss. Loyola
      of Chicago. <u>DAI</u> 35 (1974): 2276A.

**A FILM**

Sturgis, Preston, dir. <u>The Great Moment</u>. With Joel
      McCrea and Betty Field. Universal, 1944.

**INTERVIEWS**

Senecour, Edwina. Personal interview. 4 Oct. 1994.

Namath, Joe. Telephone interview. 16 Nov. 1994.

**A TELEVISION PROGRAM**

God Bless America and Poland, Too. Writ. Marian
     Marzyski. The American Experience. PBS.
     WMFE, Orlando. 10 Jan. 1991.

**A VIDEOTAPE**

The Remarkable Bandicoots. Videocassette. Dir. David
     Moore. Pacific Arts Video, 1985. 50 min.

**A PUBLICATION ON DISKETTE**

Carroll, Lewis. The Complete Annotated Alice.
     Diskette. Santa Monica: Voyager, 1991.

**AN ARTICLE ACCESSED FROM A PERIODICALLY PUBLISHED DATABASE ON CD-ROM**

Delancy, Joan. "10 Danger Signs to Look for When
     Buying a Franchise." Black Enterprise 30 Sept.
     1994: 118. Ethnic NewsWatch. CD-ROM. Softline
     Information. Feb. 1995.

NOTE:   The article was printed in the publication *Black Enterprise*.
The title of the database (underlined) is *Ethnic NewsWatch*. The
entry ends with the name of the vendor and the electronic publication date.

**A NONPERIODICAL PUBLICATION ON CD-ROM**

Impressionism and Its Sources. Windows. CD-ROM.
     Union City: Ebook, 1992.

NOTE:   No author was given in the source. "Windows" specifies the
version.

MATERIAL ACCESSED THROUGH A COMPUTER SERVICE OR NETWORK

Merkel, Jayne. "Neoclassicism at the Metropolitan."
    <u>Architecture</u> Feb. 1994: 39. <u>Architecture DB</u>.
    Online. Dialog. 24 Mar. 1995.

NOTE: *Architecture DB* is the title of the database. Following the word *Online* is the name of the computer service. The entry ends with the date of access.

## ■ Exercise 1

*Using the MLA style, make up a list of Works Cited from the following information. Place entries in the correct form and the correct order. Omit unnecessary information. Answers are on p. 393.*

A work in three volumes entitled *Mark Twain: A Biography* by Albert Bigelow Paine, published by Harper and Brothers in 1912 in New York City.

An essay in a collection of essays entitled *The Arts and the American Home, 1890–1930,* published by the University of Tennessee Press in Knoxville, Tennessee, in 1994. The essay is entitled "The Piano in the American Home" and is by Craig H. Roell. The volume is edited by Jessica H. Foy and Karal Ann Marling. The essay is on pages 85–110.

A book published in London, England, in 1971 by Longman with the title *The Prophet Harris* by Gordon MacKay Haliburton.

An article in a scholarly journal issued quarterly, the *Journal of Social Philosophy,* published in 1993 on pages 233 through 242 of volume 24, No. 3. The author's name is Victoria Davion. The title is "The Ethics of Self-Corruption."

The second edition of a book entitled *The Literature of Jazz: A Critical Guide* by Donald Kennington and Danny L. Read. The place of publication is Chicago, the date of the second edition is 1980, and the publisher is the American Library Association.

A book from the Oxford University Press of Oxford, England, by Luh Ketut Suryani and Gordon D. Jensen, published in 1993, and entitled *Trance and Possession in Bali*.

A *New York Times* newspaper article written by Malcolm W. Browne appearing in section "c" on page 1 and entitled "Perfect Violin: Does Artistry or Physics Hold Secret?" The date of the newspaper is June 14, 1994, and it is the late edition.

An article in a weekly periodical called *The Nation* with the title "Free Speech in the Internet" by Jon Wiener, appearing on pages 825 through 828 of this magazine on June 13, 1994.

No author is given for this newspaper article with the heading "Chinese Cooperation in U.N. Efforts Vowed" in the June 13th issue of the *Miami Herald* in 1994 and on page 6 of the "A" section.

One specific volume, volume 6, of a multivolume work called *Religious Trends in English Poetry* by Hoxie Neale Fairchild, published by the Columbia University Press in New York, New York, in 1968.

■ **Answers for Exercise 1, p. 392.**

Works Cited

Browne, Malcolm.  "Perfect Violin: Does Artistry or

Physics Hold Secret?"  New York Times 14 June

1994, late ed.: 1C.

"Chinese Cooperation in U.N. Efforts Vowed."  Miami

Herald 13 June 1994: 6A.

Davion, Victoria.  "The Ethics of Self-Corruption."

Journal of Social Philosophy 24 (1993): 233-42.

Fairchild, Hoxie Neale.  Religious Trends in English

    Poetry.  Vol. 6.  New York: Columbia UP, 1968.

Haliburton, Gordon MacKay.  The Prophet Harris.

    London: Longman, 1971.

Kennington, Donald, and Danny L. Read.  The Literature

    of Jazz: A Critical Guide.  2nd ed.  Chicago:

    American Library Assn., 1980.

Paine, Albert Bigelow.  Mark Twain: A Biography.

    3 vols.  New York: Harper, 1912.

Roell, Craig H.  "The Piano in the American Home."

    The Arts and the American Home, 1890-1930.

    Ed. Jessica H. Foy and Karal Ann Marling.

    Knoxville: U of Tennessee P, 1994.  85-110.

Suryani, Luh Ketut, and Gordon D. Jensen.  Trance and

    Possession in Bali.  Oxford: Oxford UP, 1993.

Wiener, Jon.  "Free Speech in the Internet."  The

    Nation 13 June 1994: 825-28.

■ **Exercise 2**

*Follow the instructions for Exercise 1. Answers are on p. 395.*

A book from Cambridge, Massachusetts, with the date 1994 and with the publisher listed as the Harvard University Press and entitled *Darwin Machines and the Nature of Knowledge*. The author is Henry Plotkin.

An article by Ellen J. Langer entitled "A Mindful Education" published on pages 42 through 50 of a quarterly periodical, the *Educational Psychologist,* in volume XXVIII in issue number one in the winter of 1993.

A book of travel by Collier, Sargent F. It is entitled *Down East: Maine, Prince Edward Island, Nova Scotia, and the Gaspé* and is published in Boston by Houghton Mifflin Company with the date of 1953.

A United States Government monthly publication, the *Federal Reserve Bulletin*—an article therein entitled "Recent Trends in the Mutual Fund Industry" by Phillip R. Mack. The date is November of 1993, and the article appears on pages 1001–1012.

A Simon and Schuster book by Louise Levathes with the title *When China Rules the Seas* and the subtitle *The Treasure Fleet of the Dragon Throne, 1405–1433.* This book was published in 1994 in New York, New York.

An article in a monthly magazine, *The American Spectator,* by Kenneth S. Lynn, entitled "Dwight Stuff," on pages 56 through 58 of the June issue in 1994.

■ **Answers for Exercise 2, p. 394.**

Works Cited

Collier, Sargent F.  <u>Down East: Maine, Prince Edward

    Island, Nova Scotia, and the Gaspé</u>.  Boston:

    Houghton, 1953.

Langer, Ellen J.  "A Mindful Education."  <u>Educational

    Psychologist</u> 28 (1993): 42-50.

Levathes, Louise.  When China Ruled the Seas:  The

    Treasure Fleet of the Dragon Throne, 1405-1433.

    New York: Simon, 1994.

Lynn, Kenneth S.  "Dwight Stuff."  The American

    Spectator June 1994: 56-58.

Mack, Phillip R.  "Recent Trends in the Mutual Fund

    Industry."  Federal Reserve Bulletin Nov. 1993:

    1001-12.

Plotkin, Henry.  Darwin Machines and the Nature of

    Knowledge.  Cambridge: Harvard UP, 1994.

### Parenthetical references

The list of Works Cited indicates clearly only what sources you used, but precisely what you derived from each entry must also be revealed at particular places in the paper. Document each idea, paraphrase, or quotation by indicating the author (or the title if the work is anonymous) and the page reference at the appropriate place in your text.

> The extent of dissension that year in Parliament has been pointed out before (Levenson 127). *[Writer cites author and gives author's name and page number in parentheses.]*
>
> Accordingly, no "parliamentary session was without severe dissension" (Levenson 127). *[Writer quotes author and gives author's name and page number in parentheses.]*
>
> Levenson points out that "no parliamentary session was without severe dissension" (127). *[Writer names author, places quotation marks at beginning of quotation and before parenthesis, and gives only page number in parentheses.]*

These references indicate that the quotation is to be found in the work by Levenson listed in Works Cited at the end of the text of the paper.

If two authors in your sources have the same last name, give first names in your references.

> It has been suggested that the general's brother did not arrive until the following year (Frederick Johnson 235).
>
> The authenticity of the document, however, has been questioned (Edwin Johnson 15).

If two or more works by the same author are listed in Works Cited, indicate which one you are referring to by giving a short title: (Levenson, *Battles* 131). The full title listed in Works Cited is *Battles in British Parliament, 1720–1721.*

If a work cited consists of more than one volume, give the volume number as well as the page: (Hoagland 2:173–74).

Some sources—the Bible and well-known plays, for example— are cited in the text of the paper but not listed as sources in Works Cited. Use the following forms:

> . . . the soliloquy (*Hamlet* 2.2). [The numbers designate act and scene. Some instructors prefer Roman numerals: *Hamlet* II.ii.]
>
> . . . the passage (1 Kings 4.3). [That is, chapter 4, verse 3.]

For further illustrations of parenthetical references to works cited, see the model research paper that follows the MLA method, beginning on p. 401.

### Notes (endnotes, footnotes)

You will probably need few if any numbered notes (called "footnotes" when at the bottom of the page, "notes" or "endnotes" at the end of the paper) because sources are referred to in the text itself and listed after the body of the paper. Explanatory notes are used to give further information or to comment on something you have written. To include incidental information in the text itself would be to interrupt the flow of the argument or to assign undue importance to matters that add to the substance only tangentially. Make your decision as to where to place such information—in the text or in a note at the end—on the basis of its impact and direct relevance.

Explanatory notes are also useful in referring to sources other than those mentioned in the text or in commenting on sources. If you wish to list several books or articles in connection with a point you are making, it may prove awkward to include all of this information in parentheses. Therefore, a note is preferable.

The sign of a note is an Arabic numeral raised slightly above the line at the appropriate place in the text. The *MLA Handbook* recommends that notes be grouped at the end of the paper. Notes should be numbered consecutively throughout.

EXAMPLES

[1] See also Drummond, Stein, Van Patten, Southworth, and Langhorne.
*[Lists further sources. Full bibliographic information given in Works Cited at end of paper.]*

[2] After spending seventeen years in Europe, he returned to America with a new attitude toward slavery, which he had defended earlier.
*[Remark parenthetical to main argument but important enough to include in a note.]*

[3] This biography, once considered standard, is shown to be unreliable by recent discoveries.
*[Evaluates a source.]*

[4] Detailed census records are available only for the 1850–1880 period. Earlier censuses do not provide the exact names of inhabitants nor any financial or social information. Later census schedules that do have detailed information on individuals are not available to the public; one may obtain only summaries of the data gathered.
*[Provides background on a source.]*

[5] Andrews uses the term *koan* to mean any riddle. To prevent confusion, however, *koan* will be used in this paper to designate only riddles in the form of paradoxes employed in Zen Buddhism as aids to meditation.

*[Clarifies terminology.]*

For further illustrations of notes, see the model research paper, pp. 401–447.

## 43i   Model research paper, MLA style

A model research paper using the MLA method of documentation is given on the following pages along with an outline and explanatory notes.

PREPARATION OF MANUSCRIPT
>   Allow ample and even margins.
>
>   Indent five spaces for paragraphs.
>
>   Double-space throughout the entire manuscript.

A title page is optional; follow the preference of your instructor. Balance the material on the page. Center the title and place it about one-third of the way down. Include your name, the name of the course (with section number), and the date.

Illusions in the Mail: Magazine Sweepstakes

By Eric Ling

English 101 (9a)

October 11, 1994

The outline should occupy a separate, unnumbered page following the title page and should follow the form illustrated on pp. 290–291.

Place the thesis statement between the title and the first line of the outline.

Illusions in the Mail: Magazine Sweepstakes

Thesis: The letters and colorful materials that consumers receive in the mail inviting them to enter magazine sweepstakes are designed to produce illusions in the service of deception.

I. Importance of illusion in selling

II. Motivations of sweepstakes promoters

III. Nature of recipients of sweepstakes materials

    A. Realists and cynics

    B. Dreamers

IV. Eight illusions of magazine sweepstakes

    A. Illusion of officiality

        1. Outer envelope

        2. Materials inside

    B. Illusion of specialness

        1. Role of computers in creating

        2. Repetition of consumer's name

    C. Illusion of probability

        1. Announcing multiple prizes

        2. Suggesting addressee may have already won

        3. Creating category of finalists

D. Illusion of advantage

    1. Purchase not required by law

    2. Widespread feeling that purchase aids chances

E. Illusion of gaming

F. Illusion of urgency

G. Illusion of celebrity

H. Illusion of philanthropy

    1. Difficulties in establishing this illusion

        a. Association with gambling

        b. Shady reputation of magazine sellers

        c. Early legal problems

    2. Cultivating image of giver to the needy

V. Magazine sweepstakes: diversion or deception?

Place page numbers in the upper right-hand corner, two lines above the first line of text. Use Arabic numerals (1, 2, 3); do not place a period after the number. Add your name before the number on each page if desired. Center your title on the page. Double-space to the first line of the text.

The author goes immediately to the subject by presenting the thesis of the paper as the first sentence. Avoid broad generalities or historical commonplaces in the opening.

The first five paragraphs, the introduction, explain and develop the thesis. The writer first establishes that sweepstakes designers are illusionists, somewhat similar to magicians. He then makes the essential point that these illusions are not created merely to sell but also to convince consumers to enter the sweepstakes. The final three paragraphs of the introduction define the difference between those people who generally refuse to take part in magazine sweepstakes and those who usually enter them, stressing their tendency to be dreamers and thus susceptible to the illusions created for them.

The parenthetical reference to Feinman, Blashek, and McCabe toward the end of the first paragraph illustrates the proper way to document a reference to three authors. If the work has more than three authors, use "et al." after the name of the first one in place of the other names (Feinman et al. 129). Notice that Feinman is the author of the quotation quoted just below. Therefore, the reference to Feinman, Blashek, and McCabe distinguishes this work from the next one cited, referred to merely as Feinman.

In order to stress the word *legerdemain,* the writer has underlined it although it was not printed in italics in the source. This change is permissible as long as the words "emphasis added" are supplied in parentheses after the page number and a semicolon.

Illusions in the Mail: Magazine Sweepstakes

The letters and colorful materials that consumers
receive in the mail inviting them to enter magazine
sweepstakes are designed to produce a series of illusions.  Promoters of sweepstakes like those administered by Publishers Clearing House, Reader's Digest,
and American Family Publishers realize the importance
of illusion in the psychology of selling.  They rely
no less upon illusion than do skilled magicians.  In
fact, toward the end of a book praising sweepstakes
for their marketing effectiveness and instructing
firms in the techniques of administering them, the
authors make the following admission:  "This has been
just a peek under the flap of the circus tent--a
glimpse of the legerdemain employed by prize promotion
specialists" (Feinman, Blashek, and McCabe 129;
emphasis added).  With magazine sweepstakes as with
magic, success depends largely upon sleight of hand.

Companies like American Family Publishers
and Publishers Clearing House do not employ sweepstakes merely to sell magazine subscriptions.  They

Prose quotations of five lines or more should be set off ten spaces as indicated here. Leave the right margin unchanged. *Do not enclose in quotation marks quotations that are set off.* Quotations within the block quotation take double (rather than single) quotation marks. See the word "No" at the beginning. MLA style calls for double-spacing of quotations that are set off, the practice followed throughout this model paper. (See pp. 160–161.) Do not further indent the first line unless more than one paragraph is being quoted.

In the first line of the block (set-off) quotation, the writer added some words of his own to clarify and explain. Such material added to quotations is called "interpolations" and must be enclosed in square brackets (*not* in parentheses).

Three spaced periods (ellipsis points) are used within a sentence to show that words have been omitted. Failure to indicate such omitted material constitutes misquotation.

The reference to Feinman at the end of the block quotation illustrates the most frequently used form of documentation: author's last name, no punctuation, page number. Note that after a block quotation, the period comes before the parenthetical documentation rather than after.

Ling 2

wish to do that, of course; it is their primary
motivation. But they also want as many people as possible to enter the sweepstakes competition even when
they decide not to buy:

> "No" names [those people who enter sweep-
> stakes but do not buy the products] are . . .
> a good revenue source. List managers will
> usually put up "no" names on a special list.
> For some offers, they are better than buyers.
> What we know about "no" names is that they
> are interested in something for nothing,
> read their direct mail, and are willing to
> spend money on a stamp for a chance to win.
> The result is an excellent list for certain
> types of insurance offers, opportunity-
> seeking plans, good luck jewelry, and of
> course, puzzle and contest clubs. Many
> mailers earn significant revenue from rental
> of "no" names. (Feinman 87)

After a quotation that has been set off, you do not necessarily have to begin a new paragraph. Here the same paragraph continues, and "Therefore" is not indented.

The reference to Clotfelter and Cook shows the proper form when a work cited has two authors. Though only one word *(infinitesimal)* is quoted, documentation is needed because that word reveals the very small chance of winning.

When an author of a quotation is mentioned by name in the text, only the page number of the quotation is needed in parentheses. When the article quoted consists of a single page, as does that of Mary Schmich, no documentation beyond the name of the author is required in the text.

Therefore, companies have a chance to make money off customers who just enter the sweepstakes--a substantial motivation.

Not everyone is interested in sweepstakes, however. Occasionally realists and cynics will enter, but as a rule they have little difficulty disposing of the bulky envelopes unopened. They realize that they have but an "infinitesimal" chance of winning (Clotfelter and Cook 46), and they are generally unwilling to take the time to perform the necessary and sometimes complicated functions to enter. The sweepstakes promotion experts do not have much success in attracting the attention of these consumers unless they just happen to want to subscribe to certain magazines anyway.

Members of another segment of the population, however, are more susceptible to the magnetic appeal of the marketing magicians. These are the customers whom Mary Schmich terms "dreamers": "When the dreamers pluck a sweepstakes announcement from the mailbox, they see possibility unfurl like a magic carpet that

Place numbers for explanatory notes slightly above the line of type and after marks of punctuation. Do not place a period after the number. Number notes consecutively throughout the paper.

The reference to J.A.T. shows documentation for a work that gives only the author's initials.

The two words quoted, *dream button,* do not need to be documented because they clearly are taken from the quotation earlier in the paragraph already documented.

The body of the essay begins with the sixth paragraph: "Because the American public. . . ." The writer is to develop in this main part of the paper the idea that sweepstakes materials create eight illusions to attract consumers, to entice them to enter, and perhaps to convince them that they should buy. In this and the next paragraph, he deals with the first illusion, that of officiality.

Maintain a level of diction consistent with Standard English. Avoid slang and informality. Formal writing, however, does not have to be stilted and dull. A phrase like *junk mail,* which is Standard English, is at the same time colorful and richly connotative.

stretches to nirvana. They scamper down that carpet toward seaside villas, red Jaguars, Mexican vacations." Judging from the number of entries in major sweepstakes, a sizable portion of the population is able to put aside realistic expectations and cynicism long enough to dream the dream of wealth and high living.[1]

The illusionists count on such a dream. Addressing them, Jeffrey Feinman advises: "Consumers enter for the chance to win a dream. To the degree that you can push that 'dream button,' you'll gain response" (88). The author of an article entitled "Sweepstakes Fever" quotes the senior vice president of a large corporation sponsoring sweepstakes: "People are developing a pot-of-gold-at-the-end-of-the-rainbow mentality" (J.A.T. 164). The sweepstakes administered by Time Life is entitled the "Million Dollar Dream." To push the "dream button," sweepstakes designers conjure up illusions.

Because the American public has developed an intense aversion to junk mail, getting people merely to

No page number in parentheses is needed for the Horowitz quotation because, as the entry under his name on the list of Works Cited indicates, the article quoted consists of only one page.

open the envelope of a sweepstakes announcement can be
a considerable challenge. To meet this challenge,
prize promoters create what might be termed the
illusion of officiality. The outer envelope will
often proclaim that important papers are enclosed and
may contain an important message to the "Postmaster"
requesting special processing in accordance with certain postal regulations. Though such announcements
mystify, they also create an illusion of officiality,
perhaps the foot-in-the-door for magazine-selling
through the mails.

The materials inside the envelope project the
same illusion with official-looking "certificates" or
"documents" that read like legal papers. With borders
that resemble those on the kinds of documents that
people generally keep in safe-deposit boxes and with
images like the American eagle that are perhaps meant
to suggest official government sanction, this sweepstakes literature makes a serious bid for attention
because "An official-looking document is hard to throw
away" (Feinman 124). Rick Horowitz comments that the

Make a dash by using two hyphens--as in this sentence--with no spaces.

The writer does not move mechanically from his discussion of one illusion to discussing another. Nothing shatters interest more than a series of paragraphs each beginning the same way as the author moves from one unit of thought to another: "The first illusion is. . . ." "The second illusion is. . . ." "The third illusion is. . . ." Though Eric Ling organized the body of his paper by his treatment of eight illusions, he did not create a dull and monotonous list.

In this research paper, the primary material consists of the sweepstakes literature itself sent through the mail. General descriptions of these advertisements and brochures do not have to be documented. The writer is simply describing what he and countless other recipients have received in the mail.

"certificate" he received in the mail had "lettering just like real money and a gold border."

Modern computers make it possible to personalize sweepstakes announcements to appeal to the individual-- repeatedly using his or her name and address. Perhaps nothing makes people feel more special than seeing their names in type or print. The computer is the best friend of those who conduct sweepstakes, for it is instrumental in creating another illusion, that of the customer's specialness, a powerful appeal to many consumers. To be addressed by one's own name is to be treated as special. It is a little like being greeted by name at the bank. Though the sweep- stakes material is sent as cheaply as possible at bulk rate, printing "personal" or "confidential" on the outside of the envelope may make it appear to be first-class mail. The illusion is enhanced as the recipient discovers that he or she is a "preferred cus- tomer" or "best customer," terms that imply special consideration.

The reference to Tooley in parentheses does not include a page number because her article is on a single page, which is given in the list of Works Cited.

The writer announces that his analysis of the illusion of probability will be subdivided into three parts, and he proceeds to discuss each aspect. Such careful attention to organization reveals an orderly mind at work and provides valuable signposts for the audience.

The reference to Feinman, Blashek, and McCabe shows how to document a quotation taken from an illustration (or photo-duplication) in a book.

Probably the most difficult obstacle that promoters of magazine sweepstakes confront is consumers' awareness of the truly slight chance they have of winning. One author estimates the "odds against winning big" at "about 150 million to 1" (Tooley). Somehow, those who design the forms for sweepstakes must be able to camouflage this discouraging reality to make entering seem worthwhile. They do this by attempting to project an illusion of probability--that is, the illusion that the person entering the sweepstakes actually has a good chance of winning--through the use of three principal methods: (1) announcing repeatedly that a large number of prizes will be awarded, (2) stating that the addressee may have already won, and (3) creating a category called "finalists."

In an announcement of its "Great American Dream Sweepstakes," Great American Magazines listed one grand prize, five first prizes, ten second prizes, and "over 53,000 additional exciting prizes" (Feinman, Blashek, and McCabe 83: illustration 30). Though many of the

When words are omitted from the end of a sentence in a quotation, represent the omission by using a period and ellipsis points (four periods, no space before the first one).

It is not necessary to repeat the same parenthetical documentation after each quotation in a series of two or three briefly separated quotations. If the information is the same for all the quotations, save it until the end of the last one—as illustrated in the passage that begins "In their instructions" and concludes with the end of that paragraph.

Ling 8

additional prizes are not worth much, the very fact
that there are so many and that the entrant is
assigned not a single prize number but several helps
to create the illusion of probability for winning. In
their instructions to potential sweepstakes promoters,
the authors of Sweepstakes, Prize Promotions, Games
and Contests suggest the following: "Blast the huge
number of prizes that are being offered. . . . You
can say, 'OVER 10,000 PRIZES GUARANTEED TO BE AWARDED!'
or 'OVER 10,000 CHANCES TO WIN.'" As an example, they
refer to the brochure of one sweepstakes that
announced: " 'OVER 88,000 PRIZES IN ALL!' This large
number was stated over and over in all the advertising. It had the power to make people feel that they
were definitely going to win a prize because there
were so many prizes available." The authors conclude
with high praise for this technique: "The result was a
tremendous number of entries" (Feinman, Blashek, and
McCabe 100).

It is common practice to suggest in the sweepstakes material that one has already won the grand

Since Daniel J. Howard and Thomas E. Barry are named in the text before the quotation, only the page number is needed in parentheses after it.

Use single quotation marks only for a quotation within a quotation: "'Winners' are relatively easy. . . ."

prize and that all that is required to collect it is
merely to send in the entry forms. To further the
illusion of probability, the recipient's name is often
included in a list of other "winners." Daniel J.
Howard and Thomas E. Barry provide a psychological
explanation for why this technique is effective:
"'Winners' are relatively easy to cajole" into turning
loose their money. In fact, creating an illusion of
winning is "a potentially powerful means of affecting
human judgment in the marketplace" (51).

A further stage in the sweepstakes process
involves notifying many people who have already
entered that they have somehow survived the first hur-
dle and are now "finalists" or "confirmed contenders."
This device allows companies another opportunity to
sell their products and widens the illusion of probability. The mere fact that so many contestants have
been eliminated seems dramatically to enhance one's
chances.

The illusion of probability merges into the illusion of advantage, a mirage created for the purpose of

The word *private* was added to the quotation for clarity. Since it is an interpolation and not part of the original quotation, it must be enclosed in square brackets.

subtly contradicting through the power of suggestion

the legal stipulation that sweepstakes not require a

consideration for entering or winning. Sweepstakes

are legal in all parts of the United States precisely

because they are not classified as lotteries, which

require some form of payment or "consideration" to

enter. Only state-operated lotteries are legal; so

"companies that run contests and sweepstakes must be

careful to avoid having them classifed as [private]

lotteries, which are illegal" (Stanley 330). There-

fore, the rules for every magazine sweepstakes state

that the consumer does not have to buy anything to

enter and win. It is highly unlikely that a large

and reputable firm would violate this law by selecting

winners only from the pool of those who order maga-

zines since the penalties for doing so would be se-

vere. "Care must be taken," admonishes an advertising

handbook, "to assure that the . . . sweepstakes is

conducted fairly and that no state or federal

regulations are violated" (McGann and Russell 312).

The consumer can rely on the claim that a person who

A paragraph should be a distinct unit of thought with a topic sentence and sentences that develop it, but unless proper transitions are supplied to connect paragraphs, an unpleasant jerkiness—even fragmentation—may result. Note how the argument flows smoothly as the writer of this paper establishes in one paragraph that major magazine sweepstakes are fair and then in the next paragraph, with the word *however* as a transition in the initial sentence, goes on to make the point that many people persist in believing they have to buy to win.

The general tone of the book by Feinman, Blashek, and McCabe is breezy and somewhat informal. Note the authors' frequent use of *it's* and other contractions. (Rosenspan, in the quotation that follows that of Feinman, Blashek, and McCabe, uses *they're*.) An academic research paper should not be breezy and informal. Feinman, Blashek, and McCabe are writing for one audience; Eric Ling is writing for another and consequently avoids the conversational informality of some of his sources (see pp. 219–220 and 292–293) while maintaining interest and clarity.

When words are omitted from the end of a sentence in a quotation, and parenthetical documentation is added, handle this in the following manner: . . ." ( ).

does not buy has just as much chance to win as one who makes a purchase.

The feeling persists among many people, however, that they have an advantage if they agree to buy. Sweepstakes promoters do all they can to sustain this illusion while at the same time clearly stating the "no consideration" rule. Rather than saying "No purchase necessary," advises a how-to book for sweepstakes creators, "it is preferable to state, 'No purchase required.' It's a stronger statement, and it subliminally suggests that entrants should purchase your product if they have any hope of winning. (Of course, there is no truth to this--it's simply a matter of reinforcing what people already believe . . .)" (Feinman, Blashek, and McCabe 105-106). According to James R. Rosenfield, "psychologically, the feeling that a purchase will somehow help you win is irresistible" (40).[2]

An expert on the "psychological appeals" of direct marketing comments that when one receives sweepstakes materials sent through the mails by

Underline for emphasis sparingly. If several words are under-lined, the effect is lost. Underlining *seem* is effective here because the writer has not made a practice of underlining and because emphasizing the word helps to call attention to its tie-in with the idea of illusion that is being developed in the paper.

Publishers Clearing House and American Family
Publishers, "you have to read their packages very
carefully to find out that they're actually selling
magazine subscriptions" (Rosenspan 367). What, then,
do they seem to be doing if not selling subscriptions?
The illusion is that they are having fun by conducting
a game that can be enormously profitable to those who
accept their appealing invitation to join in the
"play." The illusion of gaming makes the experience
of entering sweepstakes pleasurable. Many of the
tasks commonly required--placing stamps in various
positions, looking for "secret" messages, attaching
labels, and so on--are designed to involve the entrant
in the illusion that he or she is playing a sort of
game rather than being exposed to a sales pitch.

Game though it may seem, it is not a leisurely
one, for an atmosphere of urgency prevails: one must
not wait; one must not put off entering the sweepstakes; one must act immediately; time is of the
essence. Extra prizes are available for "early bird"
entrants, and all the written materials proclaim the

The word *urgent* is not underlined for emphasis as the word *seem* was previously but for another reason. Underline a word when the word *word* is used with it: The word *joy* has three letters (see **30d**).

The parenthetical page reference to Nash reads 244, 245, not 244–45, because the passage quoted does not run consecutively but occurs on the two pages indicated.

To heighten the effect of his argument about illusions, the writer occasionally uses words that recall and underscore the term employed by sweepstakes designers themselves: *legerdemain,* which is quoted in the first paragraph of the paper. Without over-working the idea, he effectively mentions "sleight of hand," refers to sweepstakes designers as "illusionists," and here calls them "magicians."

necessity of meeting absolute deadlines. The word
urgent is generously sprinkled throughout the instructions. Warnings abound. One reads: "If you fail to
return the ADDRESS LABEL below with your nine Personal
Prize Claim Numbers before . . . [the deadline date],
you would forfeit all claims to the Ten Million
Dollars even if one of your numbers is ALREADY THE
WINNING NUMBER and the entire amount would have to be
paid out to someone else" (Feinman, Blashek, and
McCabe 20: illustration 5). One marketing expert
states: "The most important of all immediacy elements
is the deadline. . . . In direct marketing, it is now
or never" (Nash 244, 245). The illusion of urgency
works to create excitement and to provoke immediate
action.

In recent years, a new illusion has been added to
those that sweepstakes magicians have traditionally
created, the illusion of celebrity. Promoters now
promise by implication not only that winners will
receive huge sums of money but also that they will be
treated like celebrities. Publishers Clearing House

Throughout the paper, the writer wisely avoids the temptation to become overtly judgmental or didactic. Though he implies that sweepstakes promoters are less than open and above board, he does not attack them directly or condemn them moralistically. Rather, he develops his argument with the objectivity and solid evidence needed to convince his audience. Reasoning is often more effective than preaching.

sends out its "Prize Patrol" with flowers and champagne

for the winner. The letter that accompanies its

"Winners Registration Portfolio" outlines an extraor-

dinary all-expenses-paid weekend that awaits the winner,

treatment that can only be described as royal--an

expensive hotel suite in New York City, limousines,

meals at gourmet restaurants--an alluring invitation

to be transported for a few moments into the lifestyle

of the rich and famous. Allowing entrants to make

selections from among a dazzling array of choices sim-

ply whets the appetite to be treated as a celebrity.

One reason for "sweepstakes fever," claims Forbes

magazine, is that "the winners are glamorized on tele-

vision" (J.A.T. 164). The leader in projecting this

illusion of celebrity is American Family Publishers,

who employ two celebrities to appear on a series of

television commercial advertisements that urge people

to enter the sweepstakes. Ed McMahon and Dick Clark,

long familiar to the American public, appear together

shortly before the deadline for entering the American

Family Publishers Sweepstakes to state in animated

Ordinarily all that is needed for parenthetical documentation is the name of the author (if not given before the quotation) and the page number. The title of the work quoted is generally not included since that information is in the list of Works Cited at the end of the paper. A title is given in the reference to Ezell because two works by the same author are referred to in the paper, and thus it is needed for clarification. See the next reference to Ezell, which is to another work. No page number is included when an article occupies a single page, which is given in the list of Works Cited.

tones that the winner will be announced on television.
A past winner usually appears with McMahon and Clark
to urge viewers to enter and to do so right away.
Depicting winners associating with Ed McMahon and Dick
Clark, the illusionists seem to be sending the message
that by winning their sweepstakes, you, too, will be a
celebrity.

So concerned with giving money away are the companies that sponsor sweepstakes that they may sometimes appear more like philanthropic institutions than
businesses selling magazines (and an ever-increasing
variety of other products) for profit. Promoters
highly prize and carefully nurture this impression--
the illusion of philanthropy--for it is difficult to
create. In the last half of the twentieth century,
"signposts marking the widening acceptability" of
sweepstakes "are plentiful" (Ezell, Fortune's Merry
Wheel 276). In former times, however, sweepstakes
were "a form of lottery" and were often condemned
as gambling (Ezell, "Sweepstakes"). They probably
originated as far back as the Middle Ages. Shakespeare

Some sources, including well-known plays like those of Shakespeare and books of the Bible, are cited in parenthetical documentation but do not have to be included in the list of Works Cited (see p. 397). The reference to *Hamlet* cites the act and scene from which the word *swoopstake* is quoted. Some writers prefer Arabic rather than Roman numerals (*Hamlet:* 4.5).

Many topics seem to call for at least a brief treatment of the historical background. The original plan for this paper called for a few lines of historical information to be presented on the first page—a seemingly natural position. Further consideration and revision, however, led the writer to move that brief section here because the long association of sweepstakes with gambling activity helps to explain why modern promoters are so intent upon trying to make sweepstakes appear respectable.

Writers must remain sensitive to the reality that an audience may be offended by terms once acceptable. Language that is gender-neutral, such as *salespersons,* is preferable to language that is male-oriented, such as *salesmen.*

Dates may be written with or without the apostrophes. See 33g.

Parenthetical documentation is not always placed immediately after the citation or quotation. Unless confusion would result, position it at the end of the sentence, as illustrated by the reference to Feinman.

uses the term (swoopstake) in Hamlet (IV.v).  In
the early eighteenth century, sweepstakes became associated with betting on horse races (Ewen 304).

The direct marketing pioneers who conceived of
using the mails to sell magazine subscriptions through
sweepstakes had both to erase the stigma of gambling
attached to this form of chance and to overcome the
somewhat shady reputation that door-to-door magazine
salespersons had earned in previous years.  Unscrupulous and obnoxious youths going door-to-door and trying to sell magazine subscriptions by claiming to be
working their way through college became a national
joke (and scandal).  But even when these unsatis-
factory techniques were replaced by direct-mail sweep
stakes in the 1950's and 1960's, all was not well.
"Most marketers were convinced that sweepstakes were
doomed" when the Federal Trade Commission investigated
sweepstakes in 1966 for failure to live up to various
laws (Feinman 85).  By agreeing to reform their proce-
dures, however, sweepstakes promoters insured a rosy
future for the industry.

Since the authors of the articles entitled "Editor's Note" and "$125,000 JPC Sweepstakes" are not known, the titles are given with the page numbers on which the quotations occur.

The writer successfully walks a fine line in his treatment of the articles in *Life* and *Ebony*. He must write nothing that could be interpreted as ridicule of the needy people who won the sweepstakes, but he must at the same time suggest that the promoters were using them, and the words that they spoke in gratitude, to make the sweepstakes appear to be an institution of kindness and generosity.

Since those days, the sweepstakes marketers have systematically worked to create for their companies the image of givers, especially to the poor. The illusion of philanthropy emerges clearly from an article published in Life magazine, the parent organization of which sponsors a major sweepstakes, which is the subject of the essay. It depicts the sweepstakes as philanthropically rewarding a deserving person, a fifty-eight-year-old woman "who grew up poor in Mississippi" and in her teens "stole bananas" to satisfy her hunger. Because of "chronic back pain," she was afraid that she might have to give up her meager paying job in a nursing home. She "had asked the Lord to see her through troubled times ahead." After she won, "she gave $5,000 to her church and paid off her debts." The final words of the article quote the winner: "My guardian angel was looking out for me" ("Editor's Note" 6). By implication, the "guardian angel" that saved this woman was the sweepstakes.

A similar attempt to project the illusion of philanthropy can be seen in Ebony, one of several

In the second and third references to the article in *Ebony,* only the page numbers are indicated in parentheses since the article was identified by title in the first reference.

The concluding section of the paper consists of the final two paragraphs. The writer first summarizes a frequently heard defense of magazine sweepstakes—that they offer a harmless diversion to millions—and then argues that although they stay within the law, they employ methods that are suspect. Here he effectively uses a series of rhetorical questions to drive home his final point.

The quotation in the paragraph beginning "Are magazine . . ." needs no parenthetical documentation since the name of the author of the article—Mary Schmich—is stated before the quotation and since the article is printed on a single page, which is included in the entry under Schmich in the list of Works Cited.

magazines owned by Johnson Publishing Company, sponsor
of a sweepstakes in 1990. The article portrays the
sweepstakes as a godsend to a poor widow, the mother
of five children. She was too ill to work and without enough money to pay her bills. "Her prayers were
answered when she won . . ." ("$125,000 JPC Sweep-
stakes" 42). She is quoted as saying, "I prayed God
would send me a miracle" (44). The sweepstakes takes
on a religious aura and seems an instrument of God
caring for the poor and needy. John H. Johnson,
Founder and Chief Executive Officer of Johnson
Publishing Company, makes his sweepstakes resemble a
philanthropic institution: "As both a widow and a
mother of five, it is a wonderful thing that she is
our major prizewinner. The level of joy . . . gener-
ated from all of our winners is nearly equaled by our
joy in bringing this special gift of happiness into
their lives" (42).

Are magazine sweepstakes, then, a harmless diversion or a base deception? The designers and promoters
of sweepstakes claim that they add excitement to

otherwise dull lives.  They argue that consumers like the elation that comes when they believe--just for a minute--that they could be fortune's favored child. They point to sentiments like that expressed by a sweepstakes lover whom Mary Schmich quotes, a feeling aroused when she learned that new laws could be passed in her state to outlaw magazine sweepstakes in their present form: "How dare they take away my dreams!  If I want reality, I can look at my checkbook."

Although sweepstakes sponsored by large companies like Publishers Clearing House, Reader's Digest, and American Family Publishers are not to be confused with dozens of crooked schemes that operate through the mails and through telemarketing to bilk consumers out of millions of dollars each year by promising rich prizes that are never delivered, they are not the benevolent institutions they purport to be either. They take great care to remain within the law and to award all the prizes just as they claim to.  But are they as ethical as they are legal?  In their reliance upon illusion, are they not manipulative and

deceptive?  Do they not pretend to be something they are not?  Are their tactics for selling not somewhat sleazy?  The very language of marketing experts who instruct businesses in designing sweepstakes at times suggests that the answers to these questions are in the affirmative: "While you don't want to humiliate consumers, you can play upon . . .[their] human weakness to achieve the results you desire--a purchase" (Feinman, Blashek, and McCabe 127).  Any activity that preys upon human weakness for profit deserves careful scrutiny.

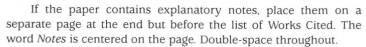

If the paper contains explanatory notes, place them on a separate page at the end but before the list of Works Cited. The word *Notes* is centered on the page. Double-space throughout.

Indent the first line of every note five spaces. Do not indent succeeding lines in a note. Numbers are raised slightly above the line.

Leave a space between the note number and the first word of the note.

As note 1 illustrates, bibliographic references in notes are identical in form to parenthetical documentation in the body of the paper.

The information in the first note supports the writer's argument that more and more people have been entering sweepstakes in recent years. Notes assist in preserving a tight structure in the body of the paper while making it possible to include materials that add further evidence and interest.

In note 2, the writer is paraphrasing a passage in Feinman, Blashek, and McCabe, whose words are given below so that the paraphrase can be compared with the source:

> This is where the yes/no psychology comes into play. Somewhere in your direct mail package you will want an involvement device that encourages prospects to buy and discourages them from just entering without buying. This device can be a perforated token or a peel-off sticker that can be affixed to the outer envelope or the response device. This is where yes/no stickers come in. Psychological research tells us that people really like to be good and please others. In essence, they like saying yes and they don't feel good when they have to say no.

Ling 21

## Notes

[1] "By the end of the go-for-broke '80s, nearly 70 percent of surveyed households had participated in sweepstakes" (Tooley).

[2] A knowledge of human psychology is essential to those who sell through magazine sweepstakes. Promoters are aware that most people are more comfortable saying "yes" than "no"; therefore, sweepstakes materials include a "Yes/No" box with instructions that entrants must affix a sticker or other object to the place that signifies whether they are buying or not. Marketers count on the natural human tendency to prefer "yes" (Feinman, Blashek, and McCabe 124–25).

Begin a new page for the list of Works Cited, which comes at the very end of the paper. Center the title. Double-space throughout.

Do not indent the first line of an entry; indent succeeding lines five spaces.

List only those sources actually mentioned in your paper and referred to in the parenthetical documentation.

Authors are listed with last names first. If a work has more than one author, list names after the first in normal order, as indicated in the first entry, which illustrates the standard form for a book with two authors.

List entries alphabetically by the authors' last names. If the author's name is not known, as in the second entry, arrange the entry according to the first word of the title (excluding *A, An,* and *The*).

Give the inclusive pages for articles as illustrated in the second entry, which shows the standard form for listing an article in a monthly magazine.

Notice that the important divisions of entries are separated by periods.

The third and fourth entries show the proper way to list a book with a single author, the one from a commercial press, the other from a university press. Subtitles are included.

Where more than one entry is necessary for an author, do not repeat the name, but type three unspaced hyphens followed by a period in place of the author's name in entries after the first. The fifth entry illustrates this form, as well as the standard way to list a signed article in an encyclopedia.

The entry for Feinman indicates the way to list an essay in a second (or later) edition of a collection with an editor. Notice that the name of the publisher (McGraw-Hill) is abbreviated to McGraw. The *MLA Handbook* supplies a list of such abbreviations of publishers' names (Section 6.5).

The three unspaced hyphens are followed by a comma if there is more than one author (Feinman, Blashek, and McCabe).

Ling 22

Works Cited

Clotfelter, Charles T., and Philip J. Cook. <u>Selling
Hope: State Lotteries in America</u>. Cambridge:
Harvard UP, 1989.

"Editor's Note." <u>Life</u> Jun. 1988: 5-6.

Ewen, C. L'Estrange. <u>Lotteries and Sweepstakes: An
Historical, Legal, and Ethical Survey of Their
Introduction, Suppression and Re-Establishment
in the British Isles</u>. London: Heath Cranton,
1932.

Ezell, John Samuel. <u>Fortune's Merry Wheel: The
Lottery in America</u>. Cambridge: Harvard UP, 1960.

---. "Sweepstakes." <u>Encyclopedia Americana</u>. 1991 ed.

Feinman, Jeffrey. "Contests and Sweepstakes." <u>The
Direct Marketing Handbook</u>. Ed. Edward L. Nash.
2nd ed. New York: McGraw, 1992. 83-92.

---, Robert D. Blashek, and Richard J. McCabe.
<u>Sweepstakes, Prize Promotions, Games and
Contests</u>. Homewood: Dow Jones-Irwin, 1986.

The entry for Horowitz illustrates the proper way to list a signed article in a daily newspaper.

The entry for Howard and Barry shows how to list an article in a scholarly journal.

When an author's initials are known but not his or her name, begin the entry with the initials, as shown in the reference to J.A.T., author of an article in a weekly magazine.

Horowitz, Rick. "This Really Takes the Prize."

Chicago Tribune 15 Jan. 1993, sec. 5: 1.

Howard, Daniel J., and Thomas E. Barry. "The Eval-

uative Consequences of Experiencing Unexpected

Favorable Events." Journal of Marketing Research

27 (1990): 51-60.

J.A.T. "Sweepstakes Fever." Forbes 3 Oct. 1988: 164-66.

McGann, Anthony F., and J. Thomas Russell. Advertising

Media: A Managerial Approach. 2nd ed. Homewood:

Irwin, 1988.

Nash, Edward L. Direct Marketing: Strategy, Planning,

Execution. New York: McGraw, 1982.

"$125,000 JPC Sweepstakes Winners." Ebony Sept. 1990:

644.

Rosenfield, James R. "While Most Consumers Think That

All Sweepstakes Are the Same, Direct Mail Prac-

titioners Know Better." Direct Marketing Oct.

1992: 38-40.

Rosenspan, Alan. "Psychological Appeals." The Direct

Marketing Handbook. Ed. Edward L. Nash. 2nd ed.

New York: McGraw, 1992. 356-69.

Schmich, Mary.   "Mail Contests Bag Only the Dreamers."

Chicago Tribune 3 Mar. 1993, sec. 2c: 1.

Stanley, Richard E.   Promotion: Advertising,

Publicity, Personal Selling, Sales Promotion.

Englewood Cliffs: Prentice, 1977.

Tooley, JoAnn.   "Lucky Numbers."   U.S. News and World

Report 21 Jan. 1991: 12.

# 43j The APA Style of Documentation

An alternate method of documentation—used widely in the sciences and the social sciences—is that explained in the *Publication Manual of the American Psychological Association* (APA), 1994.

Note that the list of works at the end of the paper is not called "Works Cited" but "References," that in the entries the date of publication comes immediately after the author's name, that initials are used instead of given names, that only the first word of a title is capitalized (except where there is a colon), that no quotation marks are used for titles, and that there are other differences from the method recommended by the Modern Language Association. Follow the preferences of your instructor as to which method you use.

## *List of references*

Include a list entitled "References" immediately after the body of the paper and before the footnotes (if any).

The APA *Manual* specifies that "references cited in text must appear in the reference list; conversely, each entry in the reference list must be cited in text," (p. 175). Alphabetically arranged, the list furnishes publication information necessary for identifying and locating all the references cited in the body of the paper. See the list for the model paper, pp. 481, 483, and study the examples below of forms used in APA documentation for various kinds of sources.

### A BOOK BY A SINGLE AUTHOR

Fine, L. A. (1990). <u>The souls of the skyscraper:
    Female clerical workers in Chicago, 1870-1930.</u>
    Philadelphia: Temple University Press.

NOTE: Use initials, not first names, of authors. Leave one space between initials. Date of publication in parentheses follows author's name. In the title use capital letters only for the first word, the first word of a subtitle, and any proper nouns. Underline titles of books.

## AN EDITED BOOK

Goodale, T. G. (Ed.). (1986). <u>Alcohol and the college student.</u> San Francisco: Jossey-Bass.

## A BOOK WITH TWO EDITORS OR AUTHORS

Luloff, A. E., & Swanson, L. E. (Eds.). (1990). <u>American rural communities.</u> Boulder, CO: Westview Press.

NOTE: Authors' last names always come first. An ampersand (&) is used before the name of the last author listed instead of the word *and*. When the name of a state is needed for clarity, use U.S. Postal Service abbreviations.

## A BOOK WITH THREE OR MORE EDITORS OR AUTHORS

Albert, E. T., Denise, T. C., & Peterfreund, H. P. (1953). <u>Great traditions in ethnics.</u> New York: American Book Company.

## A TRANSLATED BOOK

Otto, W. F. (1965). <u>Dionysus: Myth and cult.</u> (R. B. Palmer, Trans.). Bloomington: Indiana University Press. (Original work published 1933)

### A MULTIVOLUME WORK

Pederson, L., McDaniel, S. L., Adams, C., & Liao, C. (Eds.). (1989). <u>Linguistic atlas of the Gulf States</u> (Vol. 3). Athens: University of Georgia Press.

### A REPUBLISHED BOOK

Rowson, S. (1986). <u>Charlotte Temple.</u> New York: Oxford University Press. (Original work published 1794)

### A REVISED EDITION OF A BOOK

Zabel, M. D. (Ed.). (1951). <u>Literary opinion in America</u> (Rev. ed.). New York: Harper.

### SECOND OR LATER EDITION OF A BOOK

Trattner, W. I. (1989). <u>From poor law to welfare state: A history of social welfare in America</u> (4th ed.). New York: Free Press.

### AN ESSAY IN A COLLECTION OF ESSAYS OR AN ANTHOLOGY

Gongalez, G. A. (1986). Proactive efforts and selected alcohol education programs. In T. G. Goodale (Ed.), <u>Alcohol and the college student</u> (pp. 17-34). San Francisco: Jossey-Bass.

NOTE: Titles of chapters, essays in collections, and articles are not placed in quotation marks nor underlined.

**AN INTRODUCTION, FOREWORD, PREFACE, OR AFTERWORD IN A BOOK**

Caudill, H. M. (1965). Foreword. In J. E. Weller,
    Yesterday's people: Life in contemporary Appalachia
    (pp. xi-xv). Lexington: University Press of
    Kentucky.

NOTE  The word is *foreword,* not *forward.*

**MORE THAN ONE WORK BY THE SAME AUTHOR**

Mead, M. (1961). Coming of age in Samoa: A psycho-
    logical study of primitive youth for Western
    civilization. New York: Morrow.

Mead, M. (1964). Anthropology, a human science:
    Selected papers, 1939 1960. Princeton, NJ: Van
    Nostrand.

NOTE:  Works by the same author are arranged according to dates
of publication, earlier date first, as above. If works appeared in
the same year, arrange alphabetically according to the first letter
of the title (disregard *A* or *The*), and add distinguishing letters (*a, b,
c,* etc.) to dates in parentheses:

Ornstein, R. E. (1986a). Multimind. Boston: Houghton
    Mifflin.

Ornstein, R. E. (1986b). The psychology of conscious-
    ness (2nd rev. ed.). New York: Penguin Books.

**GOVERNMENT PUBLICATION, CORPORATE AUTHOR**

United States Department of Labor. (1956). The
    American workers' fact book. Washington, DC: U.S.
    Government Printing Office.

ARTICLE IN A MONTHLY PERIODICAL

Edidin, P. (1989, April). Drowning in wealth.
    Psychology Today, 23, 32-35, 74.

NOTE: Do not abbreviate names of months. Titles of articles are neither underlined nor placed in quotation marks. Discontinuous pages are listed as indicated.

ARTICLE IN A WEEKLY PERIODICAL

Ball, R. (1990, June 11). The new Elizabethans. Time,
    135, 47.

ARTICLE IN A QUARTERLY PERIODICAL WITH CONTINUOUS PAGINATION

Burns, G. (1990). The politics of ideology: The papal
    struggle with liberalism. American Journal of
    Sociology, 95, 1123-1152.

ARTICLE WITH TWO AUTHORS IN A JOURNAL PAGINATED BY ISSUE

Cannon, H. M., & Morgan, F. W. (1990). A strategic
    pricing framework. Journal of Services Marketing,
    4(2), 19-30.

NOTE: When pagination begins anew with each issue, the number of a particular issue must be given. The above entry refers to the second issue in volume 4.

ARTICLE WITH MORE THAN TWO AUTHORS IN A JOURNAL

Fouly, K. A., Bachman, L. F., & Cziko, G. A. (1990).
    The divisibility of language competence: A confir-
    matory approach. Language Learning, 40, 1-20.

**BOOK REVIEW IN A PERIODICAL**

Porte, Z. (1990). [Review of the book Family therapy
    with ethnic minorities]. Journal of Cross-Cultural
    Psychology, 21, 250-251.

NOTE: If the review has a title, place it before the material in
brackets.

**SIGNED ARTICLE IN A DAILY NEWSPAPER**

Gordon, L. (1990, May 31). Cutting costs on campus.
    Los Angeles Times, pp. A1, A38, A39.

**UNSIGNED ARTICLE (OR EDITORIAL) IN A DAILY NEWSPAPER**

Panel urges creation of immigration agency. (1990,
    June 3). Miami Herald, p. 6A.

NOTE. Most newspapers list the section number before the page
number (A6). Others (as above) may list the page number first (6A).

**DOCTORAL DISSERTATION ABSTRACTED IN *DISSERTATION ABSTRACTS
INTERNATIONAL***

Sweet, S. D. (1989). Liberal education and technology:
    A study of the process of change. Dissertation
    Abstracts International, 49, 3283A.

**FILM**

Wallis, H. B. (Producer), & Huston, J. (Director).
    (1941). The Maltese falcon. [Film]. Warner Bros.

## Reference citations in text

As you quote, summarize, or simply refer to sources in the body of your paper, identify these sources on the spot in parentheses. Give the last name of the author (or last names if more than one author) and, after a comma, the date of publication: (Golightly & Walksoftly, 1973). If you are quoting or referring to a specific part of a work, furnish the page number: (Beamish, 1901, p. 34). You can work in this information in various ways, as illustrated below. By using all these formats for parenthetical citations, you will achieve greater flexibility and thus avoid monotony.

> In motion pictures of the silver-screen era, "character development and story were paramount"; in modern films, "those essentials tend to be neglected for the less basic attributes of special effects and 'realistic' language" (Brienza, 1990, p. 183).

> Brienza (1990) observes that in the current cinema important aspects, such as characterization and plot, tend to be neglected. Several other critics (Lamond & Greely, 1983; Secord, 1989; Seltzer, 1990) feel that the art of the motion picture is currently at its highest point.

> According to Brienza (1990), "character development and story were paramount" in motion pictures of the thirties and forties, a contrast to films of our time, in which "those essentials tend to be neglected for the less basic attributes of special effects and 'realistic' language" (p. 183).

For other examples, see the model paper, pp. 463–485. In the APA style, set off (block) quotations of forty words or more. From the left margin indent five spaces; do not use quotation marks; do not indent from the right margin; place the parenthetical reference after the period ending the quotation; do not use a period after the reference. Note the following example:

Many of our struggling young people like to complain about unfairness, about how difficult it is for those without the advantages of inheritances to make it in today's world. They are not the only ones faced with inequalities, however:

> Age has inequalities that are even greater than those of youth. In the matter of income, for example, most of the old are well below the national median, but others are far above it. Many of the great American fortunes are now controlled by aged men (or by their widows or lawyers). (Cowley, 1980, p. 39)

### *Footnotes*

The APA *Manual* (p. 163) advises writers to use footnotes sparingly since all truly pertinent information should be included with parenthetical documentation in the body of the paper. Footnotes are allowed for certain purposes (most of which are not relevant to student writing assignments). You may wish, however, to include "content" footnotes, which add data or information. For an example of such a footnote with commentary, see the model paper, p. 485.

## **43k**  Model research paper, APA style

A model research paper in the APA style of documentation with accompanying explanations is given on the following pages.

Beginning with the title page, place a shortened version of your title (or the entire title if it is brief) in the upper right corner of each page with the page number in Arabic numerals. Note that in the APA style, the title page is numbered 1. Use the same heading, and number pages consecutively throughout the paper.

Center the title on the page. APA advises against using in titles "words that serve no useful purposes," such as "A Study of" or "An Experimental Investigation of" (p. 7).

After double-spacing, center your name under the title.

Follow your teacher's instructions in regard to any other information to be included on the title page. Student papers usually include, as here, the name of the course (with a section number if pertinent), the name of the professor, and the date that the paper is submitted. Double-space between all lines.

Spring Break

Alison Cohen

English 101, Professor Cedric Andressen

November 4, 1994

The abstract, usually required when following the APA format, serves as a kind of outline and thesis statement combined. It should not be more than 120 words in length. It should be carefully prepared so that it gives an accurate account of the contents and the purpose of your paper. Do not include in the abstract materials that are not in the text itself. Note: The abstract is a *summary* of the paper, not a commentary on it.

The abstract should be a separate page. Place the title of the paper (or an abbreviated form of it) in the right-hand corner of the page with the numeral 2. Double-space and center the word *Abstract*. Double-space and begin at the far left margin. Do not indent.

## Abstract

The hundreds of thousands of college students who travel each year to beaches in warm climates during that recess known as spring break have won a reputation for undesirable behavior.  Their excessive activities have been widely observed and reported with the result that spring break has come to symbolize to many something new and terrifying: the sudden decay of our youth.  A study of ancient civilizations, however, suggests that spring breaks in various forms--often called Saturnalia--have been with us since early times and have been tolerated both because they arise from natural human inclinations and because if limited and controlled they can contribute to the process of maturation.

Allow five spaces between the title (or abbreviated title) in the upper right corner and the page number. Double-space between the title of the paper (centered) and the first line of text. Leave adequate margins on the right and left as well as at the top and bottom. Indent five spaces for paragraphs.

The opening paragraph effectively describes the phenomenon known as spring break, using materials found in recent accounts. Notice that the author does not try to gloss over the less attractive aspect of the activity. If she is to be convincing, she must convey the impression of objectivity.

The title rather than an author's name is given in the first reference ("Bacchanalia or Bust") because the article is unsigned. Note that in the APA style, the date of publication is an essential part of the parenthetical reference.

The material in the sentence beginning "Wherever they go" is not quoted but paraphrased from Greene. When the author's name is given, the title is not necessary, but the date of publication is included together with the number of the page on which the material appears.

The reference to Greene & Meyer illustrates parenthetical documentation when there is more than one author.

Spring Break

Each year in March or early April hordes of
college students put aside their books and worries
about classes and grades and set out for beaches
usually in the warmer climates of the South or
California.  Much of the "better heeled college crowd
heads for Mexico, the Bahamas, or Key West"
("Bacchanalia or Bust," 1987).  Wherever they go, they
indulge themselves in various sorts of activities
including beer chugs and ketchup-drinking contests
(Greene, 1986, p. 30) while enjoying the loud and
seemingly endless sounds of rock music.  In most
colleges and universities this period of revelry is
known officially as "spring recess," but to the
students it is "spring break," an appropriate
designation since there is widespread breakage
wherever it occurs.  The title of an article in the
Chronicle of Higher Education says it all: "In Pursuit
of Alcohol and Sex, Thousands Invade Florida and 'Just
Go Nuts Down There'" (Greene & Meyer, 1986).  The

The sentence beginning "Gerlach (1989)" shows a standard way of quoting in the APA format.

The second paragraph begins with a reference ("such behavior") to the subject discussed in the first paragraph, thus offering an excellent transition to the point to be made in this paragraph: that spring break currently has an extremely bad reputation.

According to APA style, the publication date would not need to be included in connection with Gerlach if this reference had occurred in the same paragraph with the first one. Since the first reference is in the preceding paragraph, the date is given here.

The article in *Newsweek,* from which information is paraphrased here, is not signed. No page reference is furnished because all of the article appears on a single page, which is given in the list of References.

situation at locales outside Florida does not seem to differ substantially.  Gerlach (1989) observes that at increasingly popular Padre Island, Texas, "drinking is everywhere, and everything is associated with it" (p. 15).  What is "associated with it" is often property damage, especially to hotels and motels.  Automobile wrecks and violations are fairly common and arrests widespread.

Such behavior is ample proof to many that we live in an age distinctive for its moral decay.  They consider the spring break as evidence that our youth has completely separated itself from time-honored tradition and reached a point of dissolution heretofore unknown.  What other conclusions can one reach after reading accounts like those in Gerlach (1989), who comments that during spring break, "student behavior is negative in virtually every possible fashion" (p. 16)?  Riordan (1989) describes spring break as "about ten days of debauch on a beach" (p. 16), and an article in Newsweek (1986) reports

The *Publication Manual of the American Psychological Association* (1994) strongly advises against using terms like "the investigator" or "the author," awkward attempts to avoid the first-person pronoun. Such attempts can also be misleading or ambiguous. It is permissible to use "I" as does Alison Cohen here, but use it sparingly. Otherwise, the paper may seem more about yourself than the subject you have investigated.

The paragraph beginning with "I do not wish" constitutes the thesis of the paper. Having described the nature of spring break and commented on its current reputation, Alison Cohen now effectively states what it is she wishes to argue about it. In many papers the thesis statement comes earlier, often in the first paragraph. Given the method of development in this particular paper, it is appropriate in this position.

The final paragraph of this page (continuing onto the next page) begins the segment of the paper offering proof that there is a traditional aspect to spring break, that it is not a new phenomenon but closely resembles celebrations long known in civilized cultures. Alison Cohen's choice of *The Golden Bough* as a source is wise since it is one of the best-known and most reliable works on the subject of cultural patterns over many centuries and in many lands.

Since the author's name, Frazer, is mentioned before the quotation beginning with "annual period," only a page number is needed after it. Contrast this with the next quotation (beginning with "outbursts"), after which both the author's name and the

Spring Break          5

that it is not unusual for fifteen students to occupy
one room and that some have lost their lives trying to
jump from one hotel or motel balcony to another.

I do not wish to deny the excesses of my fellow
students during these vaction periods when they are
away from the restraints of home and college, but it
is important to correct the idea that spring break is
a new phenomenon (and thus represents a dire new
trend) and to understand that deplorable as some of
the behavior is, the motivation for spring break
emerges from needs in human nature that are neither
perverse nor alarming.

There is nothing new about a celebration in the
midst of busy times in which youthful exuberance is
allowed free rein.  In fact, such activity has long
been known to be part of many ancient civilizations
and cultures.  In his great work, The Golden Bough
(1913), James George Frazer devoted much attention to
this "annual period of license, when the customary
restraints of law and morality are thrown aside" and

page are given. Note that no publication date is included for Frazer here or in connection with the quotation that follows because the date for *The Golden Bough* has already been given in this paragraph.

Footnote numbers should be slightly raised above the line with no space before them. For an explanation of this footnote, see p. 484.

The paragraph beginning "The basic idea" continues the comparison between spring break and Saturnalia, emphasizing both a positive aspect, the desire for a sense of community, and a negative one, destructive vitality.

The citation to Hamilton illustrates the basic form of a parenthetical reference, giving the author's last name, date of publication, and the page number for the quotation.

when the celebrants "give themselves up to extravagant mirth and jollity, and when the darker passions find a vent which would never be allowed them in the more staid and sober course of ordinary life" (p. 306). Then, as now, matters usually got out of hand, with "outbursts of the pent-up forces of human nature, too often degenerating into wild orgies of lust and crime" (Frazer, p. 306). Just as spring break takes place toward the end of the school year, so did these traditional revelries occur during the end of the calendar year. "Now, of all these periods of license," writes Frazer, "the one which is best known and which in modern languages has given its name to the rest, is the Saturnalia" (p. 306).[1]

The basic idea behind the Saturnalia was one of fellowship and goodwill, of not only enjoying onself but also others' company. It was declared a time of peace rather than conflict, and it stressed "the idea of equality . . . when all were on the same level" (Hamilton, 1940, p. 45). No attitude seems more

The paragraph beginning with "Yet it has seemed" illustrates how a saturnalian experience can be important in the process of maturation by discussing one of Shakespeare's heroes, Prince Hal, later Henry V, King of England.

Note that the names of plays are underlined.

pervasive among the spring break crowds today than
this very spirit of camaraderie and equality.  But, as
C. L. Barber (1951) points out, "a Saturnalian
attitude, assumed by a clear-cut gesture toward
liberty, brings with it an accession of 'wanton'
vitality. . . . The energy normally occupied in
maintaining inhibition is freed for celebration"
(p. 598).  "Wanton vitality" was the main cause of
Saturnalia's getting out of hand in ancient times as
it is the reason behind such disorder and misbehavior
among spring breakers today.

Yet it has seemed important since early times
to have Saturnalia.  Why?  The answer to this question
is that such celebrations, subject as they are to becoming riotous and immoral, can nevertheless be
basically healthy because they can serve a positive
function in the total makeup of human beings.  In one
of his plays, <u>Henry IV, Part I</u>, Shakespeare depicts
his hero, Prince Hal, as enjoying one long spring
break as he rollicks in the company of disreputable

In this paper two works by the same author, C. L. Barber, are cited. No confusion results, however, because the date of publication given here, 1955, distinguishes this work from the one cited in the previous paragraph with the date 1951.

The concluding paragraph is not an unqualified defense of spring break since it suggests the need for restraint, limitations on freedom. Nevertheless, it forcefully makes the final points that the need for such celebrations appears to emerge from healthy sources within, that this motivation has long been recognized, and that spring break in some form will probably always exist.

The quotation "limit holiday" repeats Barber's words in the preceding paragraph, which was documented, and therefore needs no citation here. However, Alison Cohen has chosen to add emphasis by underlining the word *limit*. The words *italics mine* are added in parentheses to indicate that the original quotation has been altered in this way.

The two sentences beginning with "In defining Saturnalia" illustrate two ways of citing sources, the first giving the name of the author before the quotation, the second after it in parentheses.

characters.   When the time comes for him truly to grow

up, however, and to assume power as King of England,

he becomes one of that country's great rulers--better

for his long spring break with Falstaff and his band

of merrymakers.   Shakespeare "dramatizes not only

holiday but also the need for holiday and the need to

limit holiday" (Barber, 1955, p. 28).   A time of

"release," according to Barber, leads to a time of

"clarification" (p. 25).

Spring break may not be instrumental in forming

many kings or presidents these days, and there

certainly is no excuse for the total lack of

responsibility and discipline that some students

exhibit during these periods.   Many fail to "<u>limit</u>

holiday" (italics mine), but for those who do, spring

break may not be all bad either.   In defining

Saturnalia, J. E. Cirlot (1971) has written that it

embodies "a desire to concentrate into a given period

of time all the possibilities of existence" (p. 279).

It is a quest for a "way out of time" (Cirlot, p.

Spring Break          9

279). Today's spring breakers probably do not realize that they have these deep-seated motivations, but they do know that there is something about this experience like none other. The desire to participate in it has been with humankind since early civilization and is likely to be with us for a long time to come.

Begin on a new page for the list of sources called, in the APA style, *References*. Center the word on the page, double-space, and list entries in alphabetical order. Each new entry begins at the left margin; indent additional lines in individual entries five spaces.

Since the entry for "Bacchanalia or bust" lists an unsigned article, the title is given first (and the entry alphabetized in the list according to the first letters of the title). This entry shows the APA style for indicating articles in weekly magazines.

The two works by the same author, C. L. Barber, are listed according to the date of publication, the one with the earlier date first. The first entry for Barber illustrates how to list an article in a journal with continuous pagination throughout a volume. When listing a title, capitalize only the first letter of the first word, of a word coming after a colon, and of any proper noun. Do not place titles of articles in quotation marks. Underline titles of books. The second entry for Barber illustrates the form for an article or chapter in a book with an editor.

The entry for Cirlot shows how to indicate a translated work.

Alison Cohen used one volume of the several that make up *The Golden Bough* and an edition later than the first. Note the proper way of presenting this information.

## References

Bacchanalia or bust. (1987, April 18). The Economist, 300, 32.

Barber, C. L. (1951, Autumn). The saturnalian pattern in Shakespeare's comedy. Sewanee Review, 59, 593-611.

Barber, C. L. (1955). From ritual to comedy: An examination of Henry IV. In W. K. Wimsat, Jr. (Ed.), English stage comedy (pp. 22-51). New York: Columbia University Press.

Beers, busts, and tragedy. (1986, April 7). Newsweek, 107, 70.

Cirlot, J. E. (1971). A dictionary of symbols (J. Sage, Trans.) (2nd ed.). New York: Philosophical Library. (Original work published 1962)

Frazer, J. G. (1913). The golden bough: A study in magic and religion (3rd ed.). (Vol. 9). London: Macmillan.

Gerlach, J. (1989, Spring). Spring break at Padre Island: A new kind of tourism. Focus, 39, 13-16, 29

The entry for Greene & Meyer shows the form for a work with two authors.

Greene, E. (1986, April 2). At the Button in Fort

    Lauderdale. Chronicle of Higher Education, 32,

    pp. 1, 30.

Greene, E., & Meyer, T. J. (1986, April 2). In pursuit

    of alcohol and sex, thousands invade Florida and

    "just go nuts down there." Chronicle of Higher

    Education, 32, pp. 29-30.

Hamilton, E. (1940). Mythology. New York: Mentor

    Books.

Riordan, T. (1989, March 27). Miller guy life. New

    Republic, 200, 16-17.

Begin a new page for footnotes, and place it last in your paper after the list of References. Be sure you have ordered your pages in accordance with the following checklist:

1. title page
2. abstract
3. body of text
4. references
5. footnotes

The footnote here is what APA refers to as a "content footnote," one that serves to "supplement or amplify substantive information in the text" (p. 163). Though not closely enough related to the line of argument in the paper itself to warrant inclusion there, the material in this note nevertheless contributes to the interest and background of the argument. It is not always necessary to have footnotes. The APA *Manual* cautions against the inclusion of notes with "irrelevant or nonessential information" (p. 163).

Indent five spaces for each new footnote. Footnotes are numbered consecutively throughout with numerals raised slightly above the line and without periods or spaces after them.

Spring Break     12

### Footnote

[1]Writers who report on the activities of spring break often use terms like <u>rites of spring</u> and <u>ritual</u>. Greene and Meyer (1986) point out that some aspects of spring break resemble the forms of a religion (p. 31). It is not uncommon to encounter the term <u>bacchanalia</u> or <u>bacchanalian</u> in accounts of spring breaks. Saturnalia is a more inclusive term than bacchanalia and incorporates more positive elements of the experience.

# Glossary of
# Usage

# 44   Glossary of Usage   *gl/us*

Many items not listed here are covered in other sections of this book. Find them by using the index. For words found neither in this glossary nor in the index, consult a good dictionary.

**A, an**   Use *a* as an article before consonant sounds; use *an* before vowel sounds.

| | |
|---|---|
| *a* nickname | *an* office |
| *a* house [the *h* is sounded] | *an* hour [the *h* is not sounded] |
| *a* historical novel [though the British say *an*] | *an* honor |
| | *an* uncle |
| *a* union [long *u* has the consonant sound of *y*] | |

**Accept, except**   As a verb, *accept* means "to receive"; *except* means "to exclude." *Except* as a preposition also means "but."

Every legislator *except* Mr. Whelling refused to *accept* the bribe.

We will *except* (exclude) this novel from the list of those to be read.

**Accidently**   A misspelling usually caused by mispronunciation. Use *accidentally.*

**Advice, advise**   Use *advice* as a noun, *advise* as a verb.

**Affect, effect**   *Affect* is a verb meaning "to influence" or "to act upon." *Effect* is usually a noun meaning "a result," "a consequence."

The medicine did not *affect* [influence] the disease.

The drug had a drastic *effect* (consequence) on the speed of the patient's reactions.

Occasionally, however, *effect* is used as a verb meaning "to cause" or "to bring about."

The operation did *effect* [bring about] an improvement in the patient's health.

**Aggravate**   Informal in the sense of "annoy," "irritate," or "pester." Formally, it means to "make worse or more severe."

**Agree to, agree with**    *Agree to* a thing (plan, proposal); *agree with* a person.

>He *agreed to* the insertion of the plank in the platform of the party.

>He *agreed with* the senator that the plank would not gain many votes.

**Ain't**    Nonstandard or slang for *is not* or *are not*.

**All ready, already**    *All ready* means "prepared, in a state of readiness"; *already* means "before some specified time" or "previously" and describes an action that is completed.

>The riders were *all ready* to mount. [fully prepared]

>Mr. Bowman had *already* bagged his limit of quail. [action completed at time of statement]

**All together, altogether**    *All together* describes a group as acting or existing collectively; *altogether* means "wholly, entirely."

>The sprinters managed to start *all together.*

>I do not *altogether* approve of the decision.

**Allusion, illusion**    An *allusion* is a casual reference. An *illusion* is a false or misleading sight or impression.

**Alot**    Nonstandard for *a lot*.

**Alright**    Nonstandard for *all right*.

**Among, between**    *Among* is used with three or more persons or things; *between* is used with only two.

>It will be hard to choose *among* so many candidates.

>It will be hard to choose *between* the two candidates.

**Amongst**    Pretentious. Prefer *among*.

**Amount, number**    *Amount* refers to mass or quantity; *number* refers to things that may be counted.

>That is a large *number* of turtles for a pond that has such a small *amount* of water.

**An**    See **A.**

**And etc.**    See **Etc.**

**And/or**    A shortcut found chiefly in legal documents. Avoid.

**Angry at**    Informal. Prefer *angry with*.

**Anxious, eager** *Anxious* is not a synonym for *eager*. *Anxious* means "worried or distressed"; *eager* means "intensely desirous."

> The defendant was *anxious* about the outcome of the trial.

> Most people are *eager* to hear good news.

**Anyways** Prefer *anyway*.

**Anywheres** Prefer *anywhere*.

**As** Weak or confusing in the sense of *because*.

> The client collected the full amount of insurance *as* her car ran off the cliff and was totally demolished.

**At this (*or* that) point in time** Avoid. Wordy and trite.

**Awesome** Means "showing fear, wonder." Informal for *impressive*.

**Awful** Informal for *bad, shocking, ludicrous, ugly*.

**Awhile, a while** *Awhile* is an adverb; *a while* is an article and a noun.

> Stay *awhile*.

> Wait here for *a while*. [object of preposition]

**Bad, badly** See p. 80.

**Basic and fundamental** Redundant. Use one or the other.

**Because** See **Reason is because.**

**Being as, being that** Use *because* or *since*.

**Beside, besides** *Beside* means "by the side of," "next to"; *besides* means "in addition to."

> Mr. Potts was sitting *beside* the stove.

> No one was in the room *besides* Mr. Potts.

**Between** See **Among.**

**Between you and I** Wrong case. Use *between you and me.*

**Bottom line** Informal for "the basic or most important factor or consideration." Avoid in formal writing.

**Bring, take** Use *bring* to indicate movement toward the place of speaking or regarding. Use *take* for movement away from such a place.

> *Take* this coupon to the store and *bring* back a free coffee mug.

**Bug** Informal or slang in almost every sense except when used to name an insect.

**Bunch** Informal for a group of people.

**Bust, busted** Slang as forms of *burst*. *Bursted* is also unacceptable.

**Can, may** In formal English, *can* is still used to denote ability; *may,* to denote permission. Informally the two are interchangeable.

FORMAL
*May* [not *can*] I go?

**Capital, capitol** *Capitol* designates "a building that is a seat of government"; *capital* is used for all other meanings.

**Center around** Use *center in* [or *on*] or *cluster around*.

**Climactic, climatic** *Climactic* pertains to a climax; *climatic* pertains to climate.

**Compare, contrast** Do not use interchangeably. *Compare* means to look for or reveal likenesses; *contrast* treats differences.

**Compare to, compare with** In formal writing use *compare to* when referring to similarities and *compare with* when referring to both similarities and differences.

The poet *compared* a woman's beauty *to* a summer's day.

The sociologist's report *compares* the career aspirations of male undergraduates *with* those of female undergraduates.

**Complement, compliment** As a verb, *complement* means to "complete" or "go well with"; as a noun, it means "something that completes." *Compliment* as a verb means "to praise"; as a noun it means "an expression of praise."

Her delicate jewelry was an ideal *complement* to her simple but tasteful gown.

Departing guests should *compliment* a gracious hostess.

**Conscience, conscious** *Conscience,* a noun, refers to the faculty that distinguishes right from wrong. *Conscious,* an adjective, means "being physically or psychologically aware of conditions or situations."

His *conscience* dictated that he return the money.

*Conscious* of their hostility toward him, the senator left the meeting early.

**Consensus of opinion** Wordy. Prefer *consensus* or *opinion*.

**Contact**   Informal for "to write" or "to telephone."

**Continual, continuous**   Use *continual* to refer to actions that recur at intervals, *continuous* to mean "without interruption."

> The return of the bats to the lake every evening is *continual.*

> Breathing is a *continuous* function of the body.

**Contractions**   Avoid contractions *[don't, he's, they're]* in formal writing.

**Contrast**   See **Compare.**

**Cool**   Slang when used to mean "excellent" or "first-rate."

**Could care less**   Nonstandard for "could not care less."

**Could of, would of**   See **Of.**

**Couple of**   Informal for "two" or "a few."

**Criteria, phenomena, data**   Plurals. Use *criterion, phenomenon* for the singular. *Data,* however, can be considered singular or plural.

**Cute, great, lovely, wonderful**   Avoid as substitutes for more exact words of approval.

**Different from, different than**   Prefer *different from* in your writing.

> Children's taste in music is often much *different from* that of their parents.

**Differ from, differ with**   *Differ from* means "to be unlike"; *differ with* means "to disagree."

> The economic system of Cuba *differs from* that of New Zealand.

> Poland *differs with* some neighboring countries on economic policy.

**Discreet, discrete**   *Discreet* means "tactful"; *discrete* means "separate" or "distinct."

**Disinterested, uninterested**   Use *disinterested* to mean someone is "unbiased, objective"; use *uninterested* to mean someone is "not interested, indifferent."

> A *disinterested* person is helpful in solving disputes.

> *Uninterested* members of an audience seldom remember what a speaker says.

**Don't**   Contraction of *do not;* not to be used for *doesn't,* the contraction of *does not.*

**Double negative**  Avoid such nonstandard phrases as *don't have no, can't hardly,* and so on.

**Due to**  Objectionable to some when used as a prepositional phrase meaning *because of.*

OBJECTIONABLE

Due to the laughter, the speaker could not continue.

BETTER

Because of the laughter, the speaker could not continue.

**Each and every**  Redundant. Use one or the other, not both.

**Each other, one another**  When two persons are involved, use the expression *each other;* when three or more persons are involved, use the expression *one another.*

The poet and her editor wrote *each other* frequently.

Arguing heatedly with *one another,* the members of the jury deliberated the case.

**Eager**  See **Anxious.**

**Early on**  Redundant. Use *early.* Omit *on.*

**Effect**  See **Affect.**

**Elicit, illicit**  *Elicit,* a verb, means "to evoke or bring forth." *Illicit,* an adjective, means "not permitted" or "unlawful."

A pianist's innovative style will sometimes *elicit* a negative response from the critics.

Authorities discovered his participation in an *illicit* trade involving weapons.

**Emigrate, immigrate**  *To emigrate* means "to leave one's country and settle in another." *To immigrate* means "to enter another country and reside there."

Luis *emigrated from* Spain to the United States when he was only five years old.

My grandmother *immigrated to* this country from the city of Messina, Sicily, in 1919.

**Eminent, imminent**  *Eminent* means "distinguished"; *imminent* means "about to happen."

**Enthused**  Use *enthusiastic* in formal writing.

**Equally as good**  Redundant. Use *equal to* or *as good as.*

**Etc.**   Not appropriate in formal writing.

**Ever, every**   Use *every* in *everybody, every day, every now and then, every other;* use *ever* in *ever and anon, ever so humble.*

**Exam**   Considered informal by some authorities. *Examination* is always correct.

**Except**   See **Accept.**

**Expect**   Informal for *believe, suppose, suspect, think,* and so forth.

**Explicit, implicit**   *Explicit* means "directly expressed or clearly defined." *Implicit* means "implied or understood."

> Hitler confronted Neville Chamberlain with *explicit* demands.

> The Prime Minister did not have to answer; his disappointment was *implicit.*

**Fabulous**   Informal for *extremely pleasing.*

**Fantastic**   Informal for *extraordinarily good.*

**Farther, further**   Generally interchangeable, though many persons prefer *farther* in expressions of physical distance and *further* in expressions of time, quantity, and degree.

> My car used less gasoline and went *farther* than his.

> The second speaker went *further* into the issues than the first.

**Feel like**   Nonstandard for "feel that."

NOT

> I feel like I could have been elected.

CORRECT

> I feel that I could have been elected.

See **Like.**

**Fewer, less**   Use *fewer* to denote number; *less,* to denote amount or degree.

NOT

> *Less* people attended the seminar today than yesterday.

CORRECT

> *Fewer* people attended the seminar today than yesterday.

> Atlanta receives *less* rain than Seattle.

**Finalize**   Bureaucratic. Avoid.

**Fine**   Avoid as a substitute for a more exact word of approval or commendation.

**Firstly, secondly, thirdly**   Pretentious. Prefer *first, second, third.*

**Fix**   Informal for the noun *predicament.*

**Flunk**   Informal: prefer *fail* or *failure* in formal usage.

**Fun**   Informal as an adjective.

NOT
> Go to Acapulco for a *fun* vacation.

**Funny**   Informal for *strange, remarkable,* or *peculiar.*

**Further**   See **Farther.**

**Good**   Incorrect as an adverb. See **10.**

**Got to**   In formal writing prefer *have to, has to,* or *must.*

> People must [not *got to*] understand that voting is a great privilege.

**Great**   Informal for "first-rate."

**Hanged, hung**   Both *hanged* and *hung* are the past tense and past participle forms of *hang,* but they are not interchangeable. *Hanged* means "executed by hanging"; *hung* means "suspended."

> In modern times criminals are seldom *hanged.*

> The decorator *hung* a reproduction of Picasso's *Portrait of a Woman* in the living room.

**Hardly**   See **Not hardly.**

**He, she**   Traditionally, *he* has been used to mean *he* or *she.* Today this usage is unacceptable. For additional discussion, alternatives, and examples, see pp. 65–66.

**He/she, his/her**   Shortcuts often used in legal writing. For more acceptable ways to avoid sexism in use of pronouns, see p. 65.

**Himself**   See **Myself.**

**Hopefully**   When *hopefully* is used as a sentence modifier in the sense "it is hoped," it is often unclear who is doing the hoping—the writer or the subject.

VAGUE
> *Hopefully,* the amendment will be approved.

CLEAR
> Most local voters hope that the amendment will be approved.

CLEAR [but with a different meaning]
> I hope the amendment will be approved.

**If, whether**  Use *if* to indicate a condition; use *whether* to specify alternatives.

> The reception will be held outdoors *if* it does not rain.

> The reception will be held indoors *whether* it rains or not.

**Illusion**  See **Allusion.**

**Impact on**  Sociological and bureaucratic jargon. Avoid *impact* as a verb.

**Imply, infer**  *Imply* means "to hint" or "suggest"; *infer* means "to draw a conclusion."

> The speaker *implied* that Mr. Dixon was guilty.

> The audience *inferred* that Mr. Dixon was guilty.

**In, into**  *Into* denotes motion from the outside to the inside; *in* denotes position [enclosure].

> The lion was *in* the cage when the trainer walked *into* the tent.

**Individual**  Avoid using for *person.*

**Infer**  See **Imply.**

**Input**  Avoid in formal writing as a substitute for "information" or "opinion."

**In regards to**  Unidiomatic: use *in regard to* or *with regard to*.

**Interact with**  Overused and vague. Avoid for more precise language.

**Interface with**  Sociological and bureaucratic jargon. Avoid.

**Into**  See **In.**

**Irregardless**  Nonstandard for *regardless.*

**Is when, is where**  Ungrammatical use of adverbial clause after a linking verb. Often misused in definitions and explanations.

NONSTANDARD
> Combustion *is when* [or *is where*] oxygen unites with other elements.

STANDARD
> Combustion occurs when oxygen unites with other elements.

> Combustion is a union of oxygen with other elements.

**Its, it's** *Its* is the possessive case of the pronoun *it; it's* is a contraction of *it is* or *it has*.

> *It's* exciting to parents when their children succeed.

> The dog eagerly chased *its* tail.

**Kind of, sort of** Informal as adverbs: use *rather, somewhat,* and so forth.

INFORMAL
> Mr. Josephson was *sort of* disgusted.

FORMAL
> Mr. Josephson was *rather* disgusted.

FORMAL [not an adverb]
> What *sort of* book is that?

**Kind of a, sort of a** Delete the *a;* use *kind of* and *sort of*.

> What *kind of* [not *kind of a*] pipe do you smoke?

**Lay, lie** See p. 40.

**Lead, led** *Lead* is an incorrect form for the past tense *led*.

**Learn, teach** *Learn* means "to acquire knowledge." *Teach* means "to impart knowledge."

> She could not *learn* how to work the problem until Mrs. Smithers *taught* her the formula.

**Less** See **Fewer.**

**Liable** See **Likely.**

**Lie** See p. 40.

**Like** A preposition. Instead of *like* as a conjunction, prefer *as, as if,* or *as though*.

PREPOSITION
> She acted *like* a novice.

CONJUNCTION
> She acted *as if* she had never been on the stage before. [correct]

> She acted *like* she had never had a date before. [informal]

Such popular expressions as "tell it like it is" derive part of their appeal from their lighthearted defiance of convention.

Do not use *like* [the verb] for *lack*.
Do not use *like* for *that* as in *feel like*.
See **Feel like.**

**Likely, liable** Use *likely* to express probability; use *liable,* which may have legal connotations, to express responsibility or obligation.

>You are *likely* to have an accident if you drive recklessly.

>Since your father owns the car, he will be *liable* for damages.

**Literally** Do not use for *figuratively.*

>She was *literally* walking on clouds. [Means that she was *actually* stepping from cloud to cloud.]

**Loose, lose** Frequently confused. *Loose* is an adjective; *lose* is a verb.

>She wore a *loose* and trailing gown.

>Speculators often *lose* their money.

**Lot of, lots of** Informal in the sense of *much, many, a great deal.*

**Lovely** See **Cute.**

**Mad** Informal when used to mean "angry."

**Man [mankind]** Objectionable to many. Prefer *humankind* or *human beings.*

**May** See **Can.**

**Media** Plural of *medium.* Do not use to refer to a single entity, such as the press or television.

**Minimize** An absolute term meaning "to make as small as possible." Therefore, do not use with qualifying words like *greatly* or *somewhat* except in informal English.

**Moral, morale** *Moral* refers to a lesson; *morale,* to mood or spirit.

**Most** Informal for *almost* in such expressions as the following:

>He is late for class *almost* [not *most*] every day.

**Myself, yourself, himself, herself, itself** These words are reflexives or intensives, not strict equivalents of *I, me, you, he, she, him, her, it.*

INTENSIVE

>I *myself* helped Father cut the wheat.

>I helped Father cut the wheat *myself.*

REFLEXIVE

>I cut *myself.*

NOT

>The elopement was known only to Sherry and *myself.*

BUT

The elopement was known only to Sherry and *me.*

NOT

Only Kay and *myself* had access to the safe.

BUT

Only Kay and *I* had access to the safe.

**Nauseous, nauseated**    Use *nauseous* to mean "causing nausea" *[a nauseous gas];* use *nauseated* to mean "suffering from nausea."

**Neat**    Overused and informal in the sense of "pleasing" or "appealing."

NOT

It would be *neat* to be able to predict the future.

**Nice**    A weak substitute for more exact words like *attractive, modest, pleasant, kind,* and so forth.

**Not hardly**    Double negative. Avoid

**Nowhere near**    Informal for "not nearly."

Mount Whitney is *nowhere near* [use *not nearly*] as tall as Everest.

**Nowheres**    Dialectal for *nowhere.*

**Nucular**    A misspelling and mispronunciation of *nuclear.*

**Number**    See **Amount.**

**Of**    Not to be used for "have."

Had he lived, the emperor could *have* [not *of*] prevented the war.

The physician would *have* [not *of*] come had he known of the illness.

**Off of**    *Off* is sufficient.

He fell *off* [not *off of*] the water tower.

**O.K., OK, okay**    **Informal.**

**Per**    Do not use for "a" or "an" in formal writing.

Guests paid fifty dollars *a* [not *per*] plate at the political dinner.

**Percent, percentage**    *Percent* refers to a specific number: 8 percent. *Percentage* is used when no number is specified: a *percentage* of the stock. The % sign after a percentage (8%) is acceptable in business and technical writing. Write out the word *percent* in formal writing. The number before percent is given in figures except at the start of a sentence.

Employees own *8 percent* of the stock.

*Eight percent* of the stock is owned by employees.

*Percent* can be singular or plural. See **7i**.
A *percentage* takes a singular or a plural verb. See **7i**.

**Phenomena** See **Criteria.**

**Photo** Informal.

**Plus** Avoid using for the conjunction *and*.

**Precede, proceed** *Precede* means "to come before." *Proceed* means "to move onward."

Freshmen often *precede* seniors on the waiting list for campus housing.

The instructor answered a few questions and then *proceeded* with the lecture.

**Principal, principle** Use *principal* to mean "the chief" or "most imporant." Use *principle* to mean "a rule" or "a truth."

The *principal* reason for her delinquency was never discussed.

The *principal* of Brookwood High School applauded.

To act without *principle* leads to delinquency.

**Prioritize** Bureaucratic jargon for "to determine priorities." Avoid.

**Provided, providing** *Provided,* the past participle of the verb *provide,* is used as a conjunction to mean "on condition that." *Providing,* the present participle of the verb *provide,* is often mistakenly used for *provided.*

The cellist will compete in the Canadian Music Festival *provided* [not *providing*] she wins the regional competition.

**Quote** A verb; prefer *quotation* as a noun.

**Raise, rise** See p. 40.

**Real** Informal or dialectal as an adverb meaning *really* or *very*.

**Reason is [was] because** Use *the reason is [was] that*. Formally, *because* should introduce an adverbial clause, not a noun clause used as a subjective complement.

NOT

The *reason* Abernathy enlisted *was because* he failed in college.

BUT

The *reason* Abernathy enlisted *was that* he failed in college.

OR
>Abernathy enlisted *because* he failed in college.

**Reference**    Avoid as verb. Prefer *refer to.*

**Relate to**    Overused and imprecise in the sense of "to understand," "to appreciate," or "to sympathize with."

NOT
>Candidates promising higher taxes are not popular; I can relate to that.

**Respectfully, respectively**    *Respectfully* means "with respect"; *respectively* means "each in the order given."

>He *respectfully* thanked the president for his diploma.

>Crossing the platform, he passed *respectively* by the speaker, the dean, and the registrar.

**Revelant**    A misspelling and mispronunciation of *relevant.*

**Scenario**    Jargon of the film industry. Avoid in the sense of "a situation" or "a circumstance."

**Sensual, sensuous**    *Sensual* connotes the gratification of bodily pleasures; *sensuous* refers favorably to what is experienced through the senses.

**Set, sit**    See p. 40.

**Shall, will**    In strictly formal English, to indicate simple futurity, *shall* is conventional in the first person [I *shall,* we *shall*]; *will,* in the second and third persons [you *will,* he *will,* they *will*]. To indicate determination, duty, or necessity, *will* is formal in the first person [I *will,* we *will*]; *shall,* in the second and third persons [you *shall,* he *shall,* they *shall*]. These distinctions are weaker than they used to be, and *will* is increasingly used in all persons.

**Should of**    See **Of.**

**So, so that**    Use *so that* instead of *so* to express intent or purpose.

>Most people use credit cards *so that* [not *so*] they can pay their bills at the end of the month.

**Sometime, some time**    *Sometime* is used adverbially to designate an indefinite point of time. *Some time* refers to a period or duration of time.

>I will see you *sometime* next week.

>I have not seen him for *some time.*

**Sort of**    See **Kind of.**

**Sort of a**   See **Kind of a.**

**Specially**   Do not use for *especially.*

> Iced tea is *especially* [not *specially*] welcome on a hot day.

**Stationary, stationery**   An adjective, *stationary* means "not moving." A noun, *stationery* refers to writing materials or office supplies.

**Super**   Informal for *excellent.*

**Sure**   Informal as an adverb for *surely, certainly.*

> INFORMAL
>
> The speaker *sure* criticized his opponent.
>
> FORMAL
>
> The speaker *certainly* criticized his opponent.

**Sure and, try and**   Use *sure to, try to.*

> Be *sure to* [not *sure and*] notice the costumes of the Hungarian folk dancers.

**Suspicion**   Avoid as a verb; use *suspect.*

**Teach**   See **Learn.**

**Than, then**   Do not use one of these words for the other. *Than,* a conjunction, is used in comparisons. *Then,* an adverb, relates to time.

> Calculus is more complicated *than* algebra.
>
> Drizzle the pizza with oil; *then* bake it at 350°.

**That, which**   When introducing a relative clause essential to the meaning of the sentence [an essential clause], use *that.* Use *which* in a relative clause that provides the sentence with additional, but not necessary, information [a nonessential clause]. Note that the nonessential clause is set off with commas.

> The driver of the car *that* ran the red light has been fined.
>
> The jacket, *which* is several years old, has large pockets and a detachable hood.

**Their, there**   Not interchangeable; *their* is the possessive of *they; there* is either an adverb meaning "in that place" or an expletive.

> *Their* dachshund is sick.
>
> *There* it is on the corner. [adverb of place]
>
> *There* is a veterinarian's office in this block. [expletive]

**These (those) kind, these (those) sort** *These [those]* is plural; *kind [sort]* is singular. Therefore use *this [that] kind, this [that] sort; these [those] kinds, these [those] sorts.*

**Thusly** Prefer *thus.*

**Toward, towards** Although both are acceptable, prefer *toward.*

**Try and** See **Sure and.**

**Uninterested** See **Disinterested.**

**Unique** Means "one of a kind"; hence it may not logically be compared. *Unique* should not be loosely used for *unusual* or *strange.*

**Use** Sometimes carelessly written for the past tense, *used.*

> Thomas Jefferson *used* [not *use*] to bathe in cold water almost every morning.

**Utilize** Bureaucratic substitute for the shorter, and more acceptable, *use.*

**Very** Overused as an intensive; avoid.

**Wait on** Unidiomatic for *wait for. Wait on* correctly means "to serve."

**Ways** Prefer *way* when designating a distance.

> NOT
> a long *ways*
>
> BUT
> a long *way*

**When, where** See **Is when.**

**Where** Do not misuse for *that.*

> I read in the newspaper *that* [not *where*] you saved a child's life.

**While** Do not use *while* in a sentence when it can be taken to mean "although."

> AMBIGUOUS
> *While* I prepared dinner, Ann did nothing.

**Who, which** *Who* refers to people; *which* refers to things.

> Some listeners are amused by a newscaster *who* cannot pronounce names correctly.
>
> Copies of the author's last novel, *which* was published in 1917, are now rare.

**Whose, who's**   *Whose* is the possessive of *who; who's* is a contraction of *who is.*

**-wise**   A suffix overused in combinations with nouns, such as *budgetwise, businesswise,* and *progresswise.*

**Wonderful**   See **Cute.**

**Would of**   See **Of.**

**You know**   Annoying, repetitious, and meaningless when used as a question inserted in a sentence.

> Avoid: He was just scared, *you know?*

# English as a Second Language (ESL)

# 45 English as a Second Language (ESL)

For those with English as a second language (ESL), this entire handbook offers a review of the basics of parts of speech, sentence construction, and usage. The checklist below (with cross-references to sections of this book) and the lists that follow will help ESL writers to deal with areas that present special problems.

## 45a ESL Checklist

Terms in **boldface type** are defined in the Glossary of Terms, pp.513–524. Other terms are defined as they are mentioned. See also the Index.

### Omissions

- Do not omit the **subject** of a sentence.

  NOT:   Checked my ticket before boarding the plane.
  SUBJECT ADDED:   *I* checked my ticket before boarding the plane.

- Do not omit the **verb** of a sentence.

  NOT:   The work in pewter both rare and beautiful.
  VERB ADDED:   The work in pewter *is* both rare and beautiful.

  Consult:   Subjects, pp. 15–16.
  Sentences, p 26.
  Verbs, pp. 6–7.
  Completeness, pp. 93–94.
  Fragments, pp. 30–31.
  Other omissions, pp. 93–94.

## Repetitions

- Do not repeat the subject (in the same **clause**) in the form of a **pronoun.**

  NOT:   The pewter vase *it* is rare, and it is also beautiful.
  SUBJECT NOT REPEATED:   The pewter vase is rare, and it is also beautiful.

- Do not repeat an **object.**

  NOT:   Freedom is what the flag stands for *it.*
  OBJECT NOT REPEATED:   Freedom is what the flag stands for. [*What* is the **object of the preposition** *for;* delete *it.*]

  Consult:   Subjects, pp. 15–16.
  Clauses, pp. 24–25.
  Pronouns. pp. 3–5
  Objects, pp. 19–20.

## Articles *(a, an, the)*

- Use the **articles** *a* or *an* with singular count (countable) nouns (persons, places, or things that can be counted).

  The room contained *a* file cabinet, *an* old floor lamp, *a* desk, and *a* chair.

- Do not use *a* or *an* with noncount (mass) nouns (persons, places, or things that are not usually counted). (See list, p. 509).

  NOT:   Rene asked for *a* money to buy *a* gas.
  BUT:   Rene asked for money to buy gas.

- Use *the* with count nouns that name specifically.

  CORRECT:   *The* desks in the room were antiques. [specific desks]
  NOT:   *The* desks are not always antiques. [desks in general]

- Generally, do not use *the* with most singular **proper nouns**.

  NOT:   *The* Alaska is still an unspoiled state.
  BUT:   Alaska is still an unspoiled state.

  Consult:   Articles, pp. 8, 114.
  Nouns, pp. 2–3.

## Verbs

- Learn **irregular verb** forms (see list, pp. 510–511).

- Learn the meanings of two-word (phrasal) verbs (see list, pp. 511–512). A two-word or phrasal verb is a verb plus a preposition or **adverb** together making up an **idiom.**

- Use the **base form of a verb** with *do, does, did, can, could, may, might, must, shall, should, will, would,* and *ought to.*

  NOT:   *Do* magazine sweepstakes really *pays* off?
  BUT:   *Do* magazine sweepstakes really *pay* off?
  NOT:   You *must allows* three weeks for delivery.
  BUT:   You *must allow* three weeks for delivery.

- Use the **past participle** with *has, have,* and *had.*

  NOT:   *I have try* to explain the situation.
  BUT:   *I have tried* to explain the situation.

- Use the **present participle** with *am, are, be, been, is, was,* and *were.*

  NOT:   Summer *is pass* quickly.
  BUT:   Summer *is passing* quickly.

  Consult:   Verbs, pp. 6–7.
  Verb forms, pp. 38–40.
  Tenses and sequence of tenses, pp. 43–47.
  Idioms, pp. 222–223.

# **45b** ESL Lists

These lists are not complete; they are merely intended to give some of the frequently encountered noncount nouns, irregular verbs, and two-word (phrasal) verbs.

## Common Noncount Nouns

NOTE: Though all of the nouns listed below are normally noncount nouns (and have no article before them), some of them are occasionally used as count nouns.

| | | |
|---|---|---|
| advice | fish | luck |
| air | flour | makeup |
| anger | fun | money |
| bacon | furniture | news |
| baggage | garlic | oxygen |
| beauty | gas | pride |
| beef | gasoline | poverty |
| bread | gold | rice |
| butter | grass | salt |
| cement | gravy | sand |
| cereal | homework | silver |
| cheese | ice cream | sleet |
| clothing | intelligence | spinach |
| coffee | jam | sugar |
| dirt | jewelry | tuition |
| employment | junk | traffic |
| entertainment | lettuce | wealth |
| equipment | lipstick | wheat |

## Irregular Verbs

This list supplements those lists of irregular verbs on pp. 38–40.

| BASE | PAST TENSE | PAST PARTICIPLE |
|------|-----------|-----------------|
| bend | bent | bent |
| bet | bet, betted | bet, betted |
| bite | bit | bitten, bit |
| bleed | bled | bled |
| breed | bred | bred |
| catch | caught | caught |
| cling | clung | clung |
| creep | crept | crept |
| cut | cut | cut |
| fall | fell | fallen |
| feed | fed | fed |
| feel | felt | felt |
| fight | fought | fought |
| find | found | found |
| flee | fled | fled |
| forget | forgot | forgotten, forgot |
| forgive | forgave | forgiven |
| have | had | had |
| hear | heard | heard |
| hide | hid | hidden |
| hit | hit | hit |
| hold | held | held |
| keep | kept | kept |
| make | made | made |
| meet | met | met |
| pay | paid | paid |
| read | read | read |
| ride | rode | ridden |
| say | said | said |
| seek | sought | sought |
| sell | sold | sold |
| send | sent | sent |
| shake | shook | shaken |
| show | showed | shown |
| sleep | slept | slept |
| slide | slid | slid |
| speak | spoke | spoken |

| BASE | PAST TENSE | PAST PARTICIPLE |
|------|-----------|-----------------|
| spend | spent | spent |
| spring | sprang, sprung | sprung |
| stand | stood | stood |
| steal | stole | stolen |
| stick | stuck | stuck |
| stink | stank | stank |
| strike | struck | struck |
| swear | swore | sworn |
| tear | tore | torn |
| tread | trod | trodden |
| weave | wove | woven |
| weep | wept | wept |
| win | won | won |
| wind | wound | wound |

## Common Two-Word (Phrasal) Verbs

ask out (invite on a date)
back up (go backward, support)
blow over (pass quietly)
blow up (explode)
break down (collapse)
break up (separate)
bring up (to rear or to initiate)
call off (cancel)
call on (visit, ask to participate)
call up (telephone)
catch up (cease to be behind)
clean up (cleanse)
come about (happen)
come across (remit, impress as)
come back (return)
come over (visit)
come up [with] (invent)
do over (do again)
drop in (visit)
drop off (slacken, take someone to)
drop out (stop attending)

fall behind (lag)
fill out (to complete, to get larger)
fill up (fill to capacity)
get along (progress)
get along [with] (relate harmoniously)
get away (escape)
get by (succeed—barely)
get into (become involved in)
get over (recover from)
get through (endure, finish)
get up (arise)
go out [with] (to accompany)
go over (review)
grow up (mature)
hand in (submit)
hand out (distribute)
hang up (end talking on phone)
help out (assist)
hold up (to delay or to rob)
hurry up (hasten)

keep on (continue)
keep up (maintain)
leave out (omit)
look into (investigate)
look out (be on the alert)
look over (survey)
look up (search for)
make sure (verify)
make up (reconcile, invent)
make up [for] (compensate)
mix up (confuse)
pick out (choose)
pin down (identify specifically)
point out (show)
put away (save, store)
put off (postpone)
put out (extinguish)
put together (assemble)

put up [with] (tolerate)
run across (discover)
run into (encounter by chance)
run out [of] (exhaust)
show up (appear)
shut off (prevent passage through)
speak up (express opinions)
talk over (discuss)
think over (ponder)
try out (to test)
turn down (reject, decrease volume)
use up (exhaust supply of)
wake up (awaken)
wear out (use as long as possible)
wrap up (wrap a package, end)

# Glossary of
# Terms

# 46 Glossary of Terms   *gl/trm*

This is by no means a complete list of terms. See "Grammar" (pp. 1–27), the index, other sections of this book, and dictionaries.

**Absolute phrase**   See **20L.**

**Abstract noun**   See **Noun.**

**Active voice**   See **Voice.**

**Adjectival**   A term describing a word or word group that modifies a noun.

**Adjective**   A word that modifies a noun or a pronoun. See pp. 8–9.

   *Her young* horse jumped over *that high* barrier for *the first* time.

**Adjective clause**   See **Dependent clause.**

**Adverb**   A word that modifies a verb, an adjective, or another adverb (see pp. 9–10).

**Adverbial clause**   See **Dependent clause.**

**Agreement**   The correspondence between words in number, gender, person, or case. A verb agrees in number and person with its subject. A pronoun must agree in number, person, and gender with its antecedent.

**Antecedent**   A word to which a pronoun refers.

   *antecedent      pronoun*
       ↓              ↓
   When the ballet *dancers* appeared, *they* were dressed in pink.

**Appositive**   A word, phrase, or clause used as a noun and placed beside another word to explain it.

   *appositive*
       ↓
   The poet *John Milton* wrote *Paradise Lost* while he was blind.

**Article**   *A* and *an* are indefinite articles; *the* is the definite article.

**Auxiliary verb**   A verb used to help another verb indicate tense, mood, or voice. Some principal auxiliaries are forms of the verbs *to be, to have,* and *to do.* See p. 7.

I *am* studying.
I *do* study.
I *shall* go there next week.
He *may* lose his job.

**Base form**    The first principal part of a verb, also called the **infinitive form** (without *to*) or the **simple form.**

**Case**    English has remnants of three cases: subjective, possessive, and objective. Nouns are inflected for case only in the possessive *(father, father's)*. An alternative way to show possession is with the "of phrase" *(of the house)*. Some pronouns, notably the personal pronouns and the relative pronoun *who,* are still fully inflected for three cases:

SUBJECTIVE OR NOMINATIVE (acting)
    I, he, she, we, they, who

POSSESSIVE (possessing)
    my (mine), your (yours), his, her (hers), its, our (ours), their (theirs), whose

OBJECTIVE (acted upon)
    me, him, her, us, them, whom

**Clause**    A group of words containing a subject and a predicate. See **Independent clause; Dependent clause.** See pp. 24–25.

**Collective noun**    A word identifying a class or a group of persons or things. See p. 3.

**Colloquialism**    A word or expression used in familiar conversation but generally inappropriate in formal writing.

**Comma splice (or comma fault)**    An error that occurs when two independent clauses are incorrectly linked by a comma with no coordinating conjunction. See pp. 32–33.

**Common noun**    See **Noun.**

**Comparative and superlative degrees**    See **10d** and **10e.**

**Complement**    A word or group of words used to complete a predicate. Predicate adjectives, subjective complement, direct objects, and indirect objects are complements. See pp. 18–20.

**Complex, compound, compound-complex sentences**    A *complex sentence* has one independent clause and at least one dependent clause. A *compound sentence* has at least two independent clauses. A *compound-complex sentence* has two or more independent clauses and one dependent clause or more. See p. 26.

**Compound subjects** See p. 16.

**Concrete noun** See **Noun.**

**Conjugation** The inflection of the forms of a verb according to person, number, tense, voice, and mood.

**Conjunctions** Words used to connect sentences or sentence parts. See also **Coordinating conjunctions, Correlative conjunctions, Subordinating conjunctions,** and p. 11.

**Conjunctive adverb** An adverb used to relate two independent clauses that are separated by a semicolon: *besides, consequently, however, moreover, then, therefore,* and so on. See **22a.**

**Connotation** A meaning suggested by a word beyond its direct meaning. See **37c.**

**Contraction** The shortening of two words combined by replacing omitted letters with an apostrophe. See **37d.**

*I've* for *I have. Isn't* for *is not.*

**Coordinate clause** See **Independent clause.** When there are two independent clauses in a compound or a compound-complex sentence, they may be called coordinate clauses.

**Coordinating conjunctions** Simple conjunctions that join sentences or parts of sentences of equal rank *(and, but, for, nor, or, so, yet).* See p. 11.

**Correlative conjunctions** Conjunctions used in pairs to join coordinate sentence elements. The most common are *either ... or, neither ... nor, not only ... but also, both ... and.*

**Dangling modifier** A modifier that is not clearly attached to a word or element in the sentence.

DANGLING MODIFIER
   Following a regimen of proper diet and exercise, Alan's weight can be controlled.

REVISED
   Following a regimen of proper diet and exercise, Alan can control his weight.

**Declension** The inflection of nouns, pronouns, and adjectives in case, number, and gender. See especially **9.**

**Degrees (of modifiers)** See **10d** and **10e.**

**Demonstrative adjective or pronoun**   A word used to point out *(this, that, these, those).*

**Denotation**   The direct and specific meaning of a word. See pp. 215–216.

**Dependent (subordinate) clause**   A group of words that contains both a subject and a predicate but that does not stand alone as a sentence. A dependent clause is frequently signaled by a subordinator *(because, since, that, what, who, which,* and so on). It always functions as adjective, adverb, or noun.

> *adjective clause*
> The tenor *who sang the aria* had just arrived from Italy.

> *noun clause*
> The critics agreed *that the young tenor had a magnificent voice.*

> *adverb clause*
> *When he sang,* even the sophisticated audience was enraptured.

**Diagramming**   Diagramming uses systems of lines and positioning of words to show the parts of a sentence and the relationships between them. Its purpose is to make understandable the way writing is put together. (See the example below.)

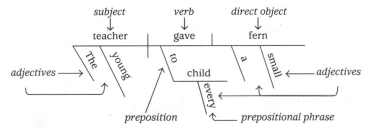

**Direct object**   A noun, pronoun, or other substantive that receives the action of the verb. See p. 19.

> The angler finally caught the old *trout.*

**Double negative**   Nonstandard use of two negative words within the same sentence.

DOUBLE NEGATIVE
> I do *not* have *hardly* any problems with my car.

REVISED
> I have *hardly* any problems with my car.

**Ellipsis points** Three spaced periods used to indicate the omission of a word or words in a quotation.

**Elliptical clause** A clause in which one or more words are omitted but understood.

> *understood*
> ↙          ↘
> The director admired no one else as much as *(he admired* or *he did)* Faith DeFelice.

**Expletive** See **7j.**

**Faulty predication** Errors occur in predication when a subject and its complement are put together in such a fashion that the sentence is illogical or unmeaningful. See **15b.**

FAULTY

> A reporter taking pictures meant that there would be publicity. [It is illogical to say that *taking pictures meant.*]

BETTER

> Since a reporter was taking pictures, we knew that there would be publicity.

**Fragment** Part of a sentence written and punctuated as a complete sentence.

FRAGMENT

> *Jostling and bantering with one another.* The team headed for the locker room.

REVISED

> Jostling and bantering with one another, the team headed for the locker room.

**Fused (or run-on) sentence** An error that occurs when two independent clauses have neither punctuation nor coordinating conjunctions between them.

FUSED

> The average adult has about twelve pints of blood this amount is roughly 9 percent of total body weight.

REVISED

> The average adult has about twelve pints of blood; this amount is roughly 9 percent of total body weight.

**Gender** The classification of nouns and pronouns into masculine, feminine, or neuter categories.

**Gerund**   See **Verbal.**

**Helping verb**   See **Auxiliary verb.**

**Idiom**   An expression peculiar to a given language. The meaning of an idiom must be memorized, for it cannot be determined simply from an understanding of the words themselves (*Tuesday week*). See **37h.**

**Imperative mood**   See **Mood.**

**Indefinite pronoun**   A pronoun not pointing out a particular person or thing. Some of the most common are *any, anybody, anyone, each, everybody, everyone, neither, one,* and *some.*

**Independent (main) clause**   A group of words that contains a subject and a predicate and that grammatically can stand alone as a sentence.

**Indicative mood**   See **Mood.**

**Indirect object**   A word that indirectly receives the action of the verb. See pp. 19–20.

   The actress wrote the *soldier* a letter.

**Infinitive**   See **Verbal.**

**Inflection**   A change in the form of a word to indicate its grammatical function. Nouns, adjectives, and pronouns are inflected by declension; verbs, by conjugation. Some inflections occur when *-s* or *-es* is added to nouns or verbs or when *'s* is added to nouns.

**Intensifier**   A modifier (such as *very*) used to lend emphasis. Use sparingly.

**Intensive pronoun**   A pronoun ending in *-self* and used for emphasis.

   The director *himself* will act the part of Hamlet.

**Interjection**   A word used to exclaim or to express a (usually strong) emotion. It has no grammatical connection within its sentence. Some common interjections are *oh, ah,* and *ouch.* See p. 14.

**Interrogative pronoun**   See p. 5 and **9f.**

**Intransitive verb**   See **Voice.**

**Inversion**   A change in normal word order, such as placing an adjective after the noun it modifies or placing the object of a verb at the beginning of a sentence.

**Irregular verb**   A verb that does not form its past tense and past participle by adding *-d* or *-ed* to its infinitive form—for example, *give, gave, given.*

**Jargon** Technical terminology or the language of a special group.

**Linking verb** A verb that does not express action but links the subject to another word that names or describes it. See pp. 7, 19, and **10b**. Common linking verbs are *be, become,* and *seem.*

**Main clause** See **Independent clause.**

**Mass noun** See **Noun.**

**Mixed construction** A sentence with two or more parts that are not grammatically compatible.

MIXED CONSTRUCTION
By cutting welfare benefits will penalize many poor families.

REVISED
Cutting welfare benefits will penalize many poor families.

**Modifier** A word (or word group) that limits or describes another word. See pp. 104–107.

**Mood** The mood (or mode) of a verb indicates whether an action is to be thought of as fact, command, wish, or condition contrary to fact. Modern English has three moods: the indicative, for ordinary statements and questions; the imperative, for commands and entreaty; and the subjunctive, for certain idiomatic expressions of wish, command, or condition contrary to fact.

INDICATIVE
*Does* she *play* the guitar?
She *does.*

IMPERATIVE
*Stay* with me.
*Let* him stay.

The imperative is formed like plural present indicative, without *-s.*

SUBJUNCTIVE
If I *were you,* I would go.
I wish he *were* going with you.
I move that the meeting *be* adjourned.
It is necessary that he *stay* absolutely quiet.
If this *be* true, no man ever loved.

The most common subjunctive forms are *were* and *be.* All others are formed like the present-tense plural form without *-s.*

**Nominal**    A term for a word or a word group that is used as a noun—for example, the *good*, the *bad*, the *ugly*.

**Nominative case**    See **Case**.

**Nonrestrictive (nonessential) modifier**    A modifier that is not essential to understanding. See **20e**.

**Nonstandard English**    See p. 212.

**Noun**    A word that names and that has gender, number, and case. There are proper nouns, which name particular people, places or things *(Thomas Jefferson, Paris,* the *Colosseum);* common nouns, which name one or more of a group *(alligator, high school, politician);* collective nouns (see **7f** and **8e**); abstract nouns, which name ideas, feelings, beliefs, and so on *(religion, justice, dislike, enthusiasm);* concrete nouns, which name things perceived through the senses *(lemon, hatchet, worm);* mass nouns (also called noncount nouns), which name things generally not counted *(silver, sugar)*.

**Noun clause**    See **Dependent clause**.

**Number**    A term to describe forms that indicate whether a word is singular or plural.

**Object**    A noun, pronoun, or word group that receives the action of the verb (**direct object**), tells to or for whom the action is done (**indirect object**), or completes the meaning of a preposition (**object of the preposition**).

**Object of preposition**    See **Object** and **Preposition, 9b**, and p. 20.

**Objective case**    See **Case**.

**Parallelism**    Parallelism occurs when corresponding parts of a sentence are similar in structure, length, and thought.

FAULTY

The staff was required to wear black shoes, red ties, and *shirts that were white*.

PARALLEL

The staff was required to wear black shoes, red ties, and white shirts.

**Participle**    See **Verbal**.

**Parts of speech**    See pp. 2–14.

**Passive voice**    See **Voice**.

**Past participle** The third principal part of a verb. Past participles generally end in *d, ed, en, n,* or *t (learned, known, dwelt).* See **3**.

**Person** Three groups of forms of pronouns (with corresponding verb inflections) used to distinguish between the speaker (first person), the person spoken to (second person), and the person spoken about (third person).

**Personal pronoun** A pronoun like *I, you, he, she, it, we, they, mine, yours, his, hers, its, ours, theirs.*

**Phrase** A group of closely related words without both a subject and a predicate. There are subject phrases *(the new drill sergeant),* verb phrases *(should have been),* verbal phrases *(climbing high mountains),* prepositional phrases *(of the novel),* appositive phrases (my brother, *the black sheep of the family*), and so forth. See pp. 21–23.

**Predicate** The verb in a clause (simple predicate) or the verb and its modifiers, complements, and objects (complete predicate). See pp. 17–18.

**Predicate adjective** An adjective following a linking verb and describing the subject. See p. 19 and **10b**.

> The rose is *artificial.*

**Predicative nominative** See **Subjective complement**.

**Predication** See **Faulty predication**.

**Preposition** A connective that joins a noun or a pronoun to the rest of a sentence. See pp. 12–13.

**Present participle** A participle formed by adding the suffix *-ing* to the base (simple) form of a verb *(searching, reading).* See pp. 22, 38–40.

**Principal parts** The verb forms **present** *(smile, go),* **past** *(smiled, went),* and **past participle** *(smiled, gone).* See pp. 38–40.

**Pronoun** A word that stands for a noun. See **Demonstrative pronoun; Indefinite pronoun; Intensive pronoun; Interrogative pronoun; Personal pronoun; Reflexive Pronoun; Relative pronoun.**

**Proper noun** See **Noun**.

**Reflexive pronoun** A pronoun ending in *-self* and indicating that the subject acts upon itself. See **Myself** (Glossary of Usage).

**Relative pronoun** See p. **5f**.

**Restrictive (essential) modifier** A modifier essential for a clear understanding of the element modified. See **Nonrestrictive (nonessential) modifier**.

**Run-on sentence** See **Fused sentence**.

**Sentence fragment** See **Fragment**.

**Sentence modifier** A word or group of words that modifies the rest of the sentence (*for example, frankly, in fact, on the other hand,* and so forth).

**Simple sentence** A sentence consisting of only one independent clause and no dependent clauses. See p. 26.

**Split infinitive** Infinitive with an element interposed between *to* and the verb form *(to **highly** appreciate)*. Avoid. See **17c**.

**Standard English** The English used in most public documents. See p. 212.

**Subject** A word or group of words about which the sentence or clause makes a statement. See pp. 15–16

**Subjective case** See **Case**.

**Subjective complement** A noun following a linking verb and naming the subject.

> The flower is a *rose*.

**Subjunctive mood** See **Mood**.

**Subordinate clause** See **Dependent clause**.

**Subordinating conjunctions** Conjunctions that connect subordinating clauses to the rest of the sentence. Some common subordinating conjunctions are *after, although, as, as if, as long as, as soon as, because, before, if, in order that, since, so that, though, unless, until, when, where, whereas, while*. See p. 11.

**Substantive** A noun or a sentence element that serves the function of a noun.

**Superlative degree** See **10d** and **10e**.

**Syntax** The grammatical ways in which words are put together to form phrases, clauses, and sentences.

**Tone** That quality in writing that conveys the author's attitude toward his or her subject.

**Transitive verb** See **Voice**.

**Verb** A word or group of words expressing action, being, or state of being. See pp. 6–7.

Automobiles *burn* gas.

What *is* life?

**Verb phrase** See **Phrase.**

**Verbal** A word derived from a verb and used as a noun, an adjective, or an adverb. A verbal may be a gerund, a participle, or an infinitive. See pp. 21–23.

**Voice** Transitive verbs have two forms to show whether their subjects act on an object (active voice) or are acted upon (passive voice). See pp. 48–49.

# Index

# Abbreviations Used in Marking Papers

*ab*    abbreviation, 167, 205
*adj*    adjective, 8, 78, 514
*adv*    adverb, 9, 79, 514
*agr*    agreement:
     pronoun, antecedent, 63, 514
     subject, verb, 52
,    apostrophe, 199
[ ]    brackets, 157
*cap*    capital letters, 200
*c*    case, 3, 71, 515
*chop*    choppy sentences, 88
*cl*    clarity, 240
:    colon, 152
,    comma, 124
*cs/fus*    comma splice, 32, 515
*no ,*    unnecessary comma, 140
*comp*    comparisons, 95
*compl*    completeness, 93
*conc*    conclusion, 261, 312
*con*    connotation, 218, 516
*cons*    consistency, 98
*coord*    coordination, excessive, 88
—    dash, 155
*dial*    dialect, 221
*d*    diction, 212
*dg*    dangling modifier, 104, 516
!    exclamation point, 168
*fig*    figurative language, 243
*fl*    flowery language, 240
*frag*    fragment, 30, 518
*fus*    fused sentence, 32, 518
*gl/trm*    glossary, terms, 513
*gl/us*    glossary, usage, 487
-    hyphen, 195
*id*    idiom, 222, 519
*intro*    introduction, 260, 311
*ital*    italics, 183
*lc*    lowercase, 201
*log*    logic, 248

*mod*    modifier, position of, 104
*mo*    mood, subjunctive, 51, 520
*ms*    manuscript form, 174
*num*    numbers, 206
*paral*    parallelism, 113, 521
( )    parentheses, 156
.    period, 161, 166
?    question mark, 162, 166
" "    quotation marks, 158
*ref*    reference, vague pronoun, 65
*rep*    repetition, 236
;    semicolon, 146
*sep*    separation of elements, 110
*sl*    slang, 221
*sp*    spelling, 187
*sub*    subordination, excessive 90
*tech*    technical diction, 225
*t*    tense, 43
*ts*    topic sentence, 262
*tr*    transition, 262, 312
*trite*    triteness, 242
*var*    variety, 117
*vf*    verb form, 38
*vocab*    vocabulary, 226
*vo*    voice, 48, 524
*w*    wordiness, 233

GENERAL EDITING MARKS

*awk*    awkward
◡    close up
   delete
✗    obvious error
∧    insert
#    insert space
○    omission
¶    paragraphing
∼    transposed